D0075527

Foundations of Plane Geometry

Foundations of Plane Geometry

Harvey I. Blau

Department of Mathematical Sciences
Northern Illinois University

PEARSON EDUCATION, INC., Upper Saddle River, NJ 07458

Library of Congress Cataloging-in-Publication Data

Blau, Harvey I.
  Foundations of plane geometry / Harvey I. Blau.
    p. cm.
  Includes bibliographical references and index.
  ISBN: 0-13-047954-3
  1. Geometry, Plane. I. Title

QA455 .B66 2003   2002030778
516.22--dc21

Editor-in-Chief: *Sally Yagan*
Acquisition Editor: *George Lobell*
Vice-President/Director of Production and Manufacturing: *David W. Riccardi*
Executive Managing Editor: *Kathleen Schiaparelli*
Senior Managing Editor: *Linda Mihatov Behrens*
Assistant Managing Editor: *Bayani Mendoza de Leon*
Production Editor: *Jeanne Audino*
Manufacturing Buyer: *Michael Bell*
Manufacturing Manager: *Trudy Pisciotti*
Marketing Assistant: *Rachel Beckman*
Art Director: *Jayne Conte*
Cover Designer: *Kiwi Design*
Cover Photo: *Palazzi di Firenze E Toscana.* © *Bencini Photographic Archive/Magnus Edizioni*
Compositor: *Eric Behr*

©2003 by Pearson Education, Inc.
Pearson Education, Inc.
Upper Saddle River, New Jersey 07458

Printed in the United States of America

10 9 8 7 6 5 4 3 2 1

ISBN 0-13-047954-3

Pearson Education LTD., *London*
Pearson Education Australia PTY, Limited, *Sydney*
Pearson Education Singapore, Pte. Ltd
Pearson Education North Asia Ltd, *Hong Kong*
Pearson Education Canada, Ltd., *Toronto*
Pearson Educación de Mexico, S.A. de C.V.
Pearson Education -- Japan, *Tokyo*
Pearson Education Malaysia, Pte. Ltd

Contents

Preface

This is a text for an upper level undergraduate course in plane geometry. It presents a unified account of the foundations of Euclidean and non-Euclidean planes. It proceeds from rather general axioms and yields a classification theorem for the three fundamental classical planes: Euclidean (or parabolic), spherical (or doubly elliptic), and hyperbolic. The treatment is careful, rigorous, and tightly focused, but it takes small and leisurely steps. I have used this approach for 15 years and have found it to be successful for our students, most of whom have been prospective secondary school mathematics teachers and have had little prior experience with abstraction and proof.

The abstract exposition is grounded in concrete examples, including the coordinate Euclidean plane, the sphere, the Beltrami-Klein hyperbolic plane, the Minkowski plane, and the "gap" plane, which are presented early (in Chapter 1) and are cited often. The frequent comparison of different models is a strong motivation for the study of the relationships among various geometric properties and of why they hold or fail in particular contexts. (See, for instance, the discussion in Chapter 1 of the Exterior Angle Inequality for the Euclidean plane and its failure for the sphere.) The diversity of examples also justifies the study of concepts such as betweenness and separation, which the student might dismiss as obvious in the context of the Euclidean plane alone. An awareness of some bizarre examples helps to motivate the introduction of axioms as a way of eliminating pathology and of homing in on the fundamental models.

Here are some remarks and suggestions about specific chapters of the book:

I think that it is important to spend a little time, but not too much, on Chapters 2, 3, and 4. Chapter 2 addresses our students' most common logical blunders and presents basic ideas about proofs. Its purpose is to enable students to understand our corrections of their logical errors throughout the semester, not to make them instant experts. (For example, if one is to find a model where a particular "If/then" statement fails, it is essential to know the general criterion for when such an implication is false.) Chapter 3 uses logical puzzles as a familiar way of gaining practice in creating and writing proofs, and Chapter 4 reviews the Least Upper Bound Property of the real numbers.

The gradual introduction of the axioms and the development of some of their consequences takes up Chapters 5–13. New concepts are defined as soon as they make sense in context, and not necessarily before all axioms that relate to them have been introduced. This allows the occurrence of strange examples. For instance, "segment" and "ray" are defined in Chapter 6, and an example (the "Inside Out" model) is given wherein segments can have more than one set of endpoints and every point of a ray can be an endpoint. By studying and constructing such examples, students begin to understand that properties of a concept are not automatic, that any particular set of postulates has its strengths and its limitations, and that taking anything for granted is not a good idea. This understanding generally takes a few weeks to form; careful guidance and a little nurturing on the part of the instructor is usually needed, particularly in traversing the material of Chapters 6 through 10.

Most of the axioms for coterminal rays are formulated (in Chapter 11) as exact analogs of previous axioms for collinear points. This analogy (duality) is invoked to establish instantly many properties of coterminal rays. The articulation of these properties helps to reinforce understanding of the previous results about collinear points and helps to justify to the student the time spent on those results. After the statement of axioms is completed in Chapter 13, the full list of assumptions is reviewed and commented on, and the ruler and protractor properties (which are theorems in our setup) are discussed.

The general theory that continues through Chapters 13–19 includes criteria for congruence of triangles, perpendicularity, the Exterior Angle Inequality (to the extent that it is true), the Triangle

Inequality, angle sums of triangles, and parallel lines. It culminates in the classification theorem (19.4) mentioned previously.

Chapters 20 and 21 study concurrence and circles, respectively, in the general context, and Chapter 22 treats similarity in a Euclidean plane. Appendix I reproduces a list of Euclid's definitions and assumptions, which are referred to several times in the text. Appendix II contains a derivation of formulas for angle measure in the Beltrami-Klein model, as well as a complete proof that this model satisfies the Side-Angle-Side congruence axiom.

I have found it possible in most semesters, with careful planning, to cover Chapters 0–19 and to treat in detail the proofs in all but the last two or three of these chapters. I have used the material in Chapters 20–22 and Appendix II for independent study projects for honors students. An exceptionally well-prepared class would be able to skip Chapters 2–4 and cover the entire book in a semester.

Whatever their prior level of preparation and mathematical maturity, I believe that most junior- and senior-level mathematics majors, particularly those who plan to teach high school mathematics, will benefit from a careful study of this book. They will gain an awareness of some rather surprising properties of hyperbolic and spherical geometry, understand better the relationships among some familiar Euclidean properties, and discover some unfamiliar Euclidean properties as well. But even more important, they will develop their abilities to understand abstract and rigorous arguments, to solve nontrivial problems, and to create and articulate reasoned and coherent proofs. Many students have told me that gaining and using such skills is a source of much enjoyment and satisfaction.

Acknowledgments

I owe a huge debt to John E. Wetzel. He was, at one time and until diverted by other commitments, a coauthor of this book. I am grateful for his many contributions.

Thanks are due to my colleagues Richard Blecksmith and Kitty Holland, who taught this material in note form several times and shared with me a lot of insightful and much-appreciated suggestions. I also wish to thank the reviewers of this text for their illuminating and helpful comments: Debra Carney, University of Connecticut; Al

Hibbard, Central College, Iowa; Robert Hunter, Penn State University; Kamal Narang, University of Alaska, Anchorage; Jack Porter, University of Kansas; and Gerard Venema, Calvin College.

The remarks and suggestions of many of my geometry students over the years were very influential in shaping the form and content of the book. I wish to thank all of them.

I thank Lisa Allison and Elizabeth Mehren for their professional and diligent work in typing most of this text. I am especially grateful to Eric Behr for expert assistance with figures, formatting, and many subtleties of technical production.

I also thank my editor, George Lobell, for his patience and help throughout the long term of this project. I thank the production staff at Prentice Hall for their aid in bringing the manuscript into a publishable form. In particular, I am grateful to the production editor, Jeanne Audino, and the copy editor, Patricia M. Daly.

Last but most important, I thank my wife Liz for her understanding, love, and support every day.

Harvey Blau
Northern Illinois University
blau@math.niu.edu

Foundations of Plane Geometry

0   The Question of Parallels

We sketch a very brief history of certain aspects of geometry as background for the material in this book. Euclid is the central figure but is neither the beginning nor the end of our story.

"Geometry" is originally a Greek word that means "measurement of the earth." The abstract concepts of shape, length, area, and their relationships have been known, at least on a practical level, from the time when human beings first began to construct buildings, mark the boundaries of agricultural fields, and chart the location and movement of the moon and stars. Some nontrivial geometric results were widely distributed as early as the Neolithic period (about 4000 B.C. to 2000 B.C.). For example, the theorem of Pythagoras (that the square of the hypotenuse of a right triangle equals the sum of the squares of the other two sides) appears to have existed then in Babylonia, the British Isles, Greece, India, China, and Egypt. The evidence for this includes the inscription of tables of Pythagorean triples (integers that are the dimensions of right triangles, such as (3, 4, 5) or (5, 12, 13)) on Babylonian baked clay tablets from the time of Hammurabi (around 1800 B.C.); configurations of right triangles of dimensions 12, 35, 37 (measured in half-integer multiples of the megalithic yard of 0.829 m) that occur in rings of standing stones near Inverness in Scotland and at Woodhenge in southern England; and an explicit statement of the theorem in the Baudhayana Sulvasutra, an ancient Hindu manual that gives precise details for the construction of altars of specified form and size.

The desire to understand why geometric facts are true, if not to prove them rigorously, also seems to be of ancient origin. A diagram and brief explanation presents an informal proof of the theorem of

1

Pythagoras in the earliest known Chinese text on astronomy and mathematics, the *Chou Pei Suan Ching* ("Arithmetical Classic of the Gnomon and the Circular Paths of Heaven"; "gnomon" means both the raised part of a sundial and a figure that equates a rectangle with a difference of squares). It is done explicitly for a triangle of dimensions 3, 4, 5 but is in fact quite general. The text was written during the Han Dynasty, around 200 B.C., but is believed to reproduce knowledge that is much older. The diagram and a hint for the proof are given in Problem 1 at the end of this chapter.

The results found on the Babylonian clay tablets, or on ancient Egyptian papyrii, are examples of primary historical evidence. That is, the artifacts were created at or near the time when (historians believe) the information written on them was discovered. There have been no such primary sources found for the more recent and more extensive contributions to geometry made by Greek civilization. The physical fragility of the Greek manuscripts and their exposure to various political and social upheavals probably accounts for this. But the study of copies and translations made hundreds of years after the fact yields a fairly clear picture of the nature and magnitude of the geometry developed in ancient Greece.

Greek mathematicians assimilated much of the geometric lore of Babylonia and Egypt. Then, from around 600 B.C., they began to apply to the subject the methods of logic developed by the Greek philosophers. In doing so they transformed geometry from a collection of empirical facts and isolated explanations into a unified discipline in which all results are proved by rigorous arguments from a small set of initial assumptions. Thales of Miletus, who lived during the sixth century B.C., is credited as being the first to use logical deduction as a standard procedure for doing geometry. About 50 years later, Pythagoras and the mystical religious brotherhood that he founded in southern Italy extended the method to connect long chains of results. The theorem that is his namesake preceded him by centuries, but Pythagoras and his school contributed many original and fundamental results. They used properties of parallel lines to investigate proportionality in similar figures and to prove that the angle sum of any triangle is equal in measure to a straight angle. One of the Pythagoreans, Hippocrates of Chios, is believed to have made the first attempt at a cohesive logical treatise on the entire field. Several other such works appeared over the next two hundred years. These efforts culminated in Euclid's *Elements*, written around 300 B.C.

Euclid served as the first Professor of Mathematics at the University of Alexandria, in Egypt at the Mediterranean Sea. The conquests of Alexander the Great made Egypt part of the Greek empire. The school at Alexandria became a major center of learning and remained so for centuries. From there, the *Elements* was disseminated through much of the world. Because of its clarity and cohesiveness, it soon became the standard text for geometry and replaced all of the works of Euclid's predecessors. Apart from some references to them in later sources, these earlier efforts no longer exist.

The initial assumptions (postulates, axioms) and definitions on which Euclid based his development are recalled in Appendix I. (Euclid's statements, both in this chapter and in Appendix I, are quoted from T. L. Heath, *The Thirteen Books of Euclid's Elements*, Vol. I, Dover Publications, New York, 1956.) Euclid's use of further tacit assumptions is discussed briefly in Chapter 13. No one saw a need to make these unspoken suppositions explicit until the late nineteenth century, when the discovery of hyperbolic geometry generated new questions about the nature of geometric truth and its verification. But the mathematical world was focused for more than 2000 years on decreasing rather than increasing the number of Euclid's assumptions. The goal was to eliminate the need for the Fifth Postulate (see Figure 0.1):

> That, if a straight line falling on two straight lines makes the interior angles on the same side less than two right angles, the two straight lines, if produced indefinitely, meet on that side on which are the angles less than the two right angles.

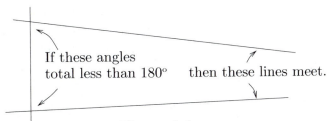

If these angles total less than 180°   then these lines meet.

Figure 0.1

This statement is about parallel (that is, nonintersecting) lines in the plane. It is equivalent, in the context of Euclid's other assumptions, to Playfair's Postulate (after the Scottish physicist and mathematician John Playfair [1748–1819]):

Through a given point not lying on a given line there can be drawn only one line parallel to the given line.

A discussion of the equivalence of these two statements, in the more general setting of our text, is given in Chapter 17. (See Theorem 17.6; see also Problems 3 and 4 at the end of this chapter.)

The properties of parallel lines and their role in plane geometry were big issues for many years before Euclid wrote the *Elements*. There is some evidence that in work from the fourth century B.C. but now lost, the philosopher Aristotle articulated as a basic assumption a principle very similar to the Fifth Postulate. In his treatise *Prior Analytics*, which still exists, he cites the implicit supposition of properties of parallel lines as an example of the logical error of begging the question.

The Fifth Postulate is both more complicated and less obvious than Euclid's other assumptions. So it was inevitable that the urge to modify it, either by proving it as a consequence of the other postulates or by replacing it with a more self-evident statement, would be reborn many times over two millennia. Only a few decades after Euclid, the great Greek mathematician Archimedes wrote a treatise (now lost) called *On Parallel Lines*. The astronomer Ptolemy of Alexandria attempted a proof of the Fifth Postulate in the second century A.D. A number of Byzantine scholars tried to construct a proof, in particular Simplicius in the sixth century. There is reason to believe that Archimedes may have redefined parallel lines as lines that are everywhere equidistant. If so, then he tacitly made an assumption that is equivalent to the Fifth Postulate. Simplicius made another such tacit assumption when he based his proof by contradiction on the "fact" that for any point in the interior of an angle, there exists a line through the point that meets both sides of the angle. This again is a property of the Euclidean plane that is equivalent to the Fifth Postulate. (See Problem 7.) Ptolemy's proof foundered on simply a gross misuse of language and logic.

All of the numerous other attempts at proving the Fifth Postulate inevitably failed. This is because there really exists another type of plane (now called hyperbolic) that is as valid as the standard Euclidean plane, and in which Euclid's other postulates hold but the fifth one fails. Both a specific model for the hyperbolic plane and its general properties are developed throughout our text, beginning in Chapter 1. It took until well into the nineteenth century before the hyperbolic plane was discovered and its existence definitively proved. We mention here only a few of the many who tried to prove the Fifth Postulate. Most of them committed an error of a type similar to one of those given in the preceding paragraph, even though the various geometers usually were rather perceptive in identifying the mistakes of their predecessors. Some were explicit in noting that they assumed an additional premise.

Mathematical leadership resided in the Arabic-speaking world from the sixth through the thirteenth centuries. Notable among those from Arabic cultures who tackled the Fifth Postulate were the Baghdad mathematician and astronomer Thabit Ibn Qurra in the ninth century and the Egyptian physicist, mathematician and astronomer Abu Ali Ibn al-Haytham in the tenth century. Both Ibn Qurra and Ibn al-Haytham introduced a quadrilateral with three right angles and presented flawed proofs that such a figure must be a rectangle. This quadrilateral was later considered in the eighteenth century by Lambert and is discussed later.

The celebrated Persian mathematician, astronomer, philosopher and poet Omar Khayyam also tried to establish the Fifth Postulate. Khayyam explicitly assumed Aristotle's principle. He also introduced the quadrilateral studied more extensively 600 years later by Saccheri and also discussed later (see as well the problems for Chapter 19).

A diverse set of mathematicians in medieval Europe attempted to prove Euclid's parallel postulate. Among them were the Polish scholar Vitello in the thirteenth century, and the Jewish religious philosopher Levi ben Gerson, who lived in southern France in the fourteenth century. John Wallis, in seventeenth-century England, adopted the postulate (equivalent to the Fifth) that there exist similar plane figures of arbitrary size. He regarded this as a natural generalization of Euclid's Third Postulate, that there are circles of arbitrary radius.

The eighteenth century Italian Jesuit priest Girolamo Saccheri was not happy with Wallis's proof. Saccheri considered quadrilaterals with two sides of equal length both perpendicular to the base, as did Khayyam. He attempted to show, as did Khayyam and others, that the two (necessarily equal) angles at the top of the quadrilateral could not be acute. He assumed the angles were acute and tried to argue toward a contradiction. This acute angle hypothesis is exactly what is true in the hyperbolic plane, but of course Saccheri was not aware of this. He was careful not to make any further assumptions, and he derived many consequences of his basic hypothesis. Thus he unknowingly developed many properties of the hyperbolic plane. In particular, he showed that any two nonintersecting straight lines either have a common perpendicular or come asymptotically close to each other. His efforts to derive a contradiction from this or consequent properties led either to an error in calculating or to the apparently frustrated declaration that the "hypothesis of acute angle is absolutely false; because (it is) repugnant to the nature of the straight line." Saccheri clearly was aware that he had not produced a satisfactory argument.

Johann Heinrich Lambert, working in Munich and Berlin in the latter half of the eighteenth century, studied the same quadrilateral with three right angles as did Ibn Qurra and Ibn al-Haytham. Lambert assumed toward a contradiction that the fourth angle was acute. Ibn al-Haytham had dismissed this situation by the same sort of re-definition of parallel lines as being everywhere equidistant that was attributed to Archimedes. But Lambert argued more carefully and derived even more properties of the hyperbolic plane than had Saccheri. In particular, he showed that the angle sum of a triangle is less than two right angles (that is, less than 180 in degree measure) and noted the resulting connection between angle sum and area. He also observed the analogy between the consequences he found of the acute angle hypothesis and the results of spherical geometry, where the fourth angle of a quadrilateral with three right angles must be obtuse. He even had the glimmer of an idea that perhaps some geometric surface, an imaginary sphere, exists where the Fifth Postulate fails and Euclid's other postulates hold. Nevertheless, he felt obliged to finish the quest for a contradiction. His argument stumbled over an unwarranted assumption about the set of points all the same distance from a given line, an assumption that resembles the one made by Ibn al-Haytham.

The Italian-French mathematician Adrien-Marie Legendre made an extended series of attempts at proving the Fifth Postulate in the 12 editions of his textbook, *Elements de Geometrie*, which appeared between 1794 and 1823. He was one of many who replicated the error made by Simplicius some 1300 years before.

The truth about the nature of the Fifth Postulate finally was grasped in the nineteenth century, by three men working independently: Nicolai Ivanovich Lobachevsky (1793–1856) of Russia, Janos Bolyai (1802–1860) of Hungary, and the great mathematician Karl Friedrich Gauss (1777–1855) of Germany. In 1829, Lobachevsky published a paper entitled "On the Principles of Geometry" where he asserted that there is a consistent system of plane geometry in which there exist multiple lines through a point that do not meet a given line and angle sums of triangles are less than the measure of a straight angle. He did not have a proof of consistency. But he did derive the trigonometric formulas that hold in his alternative plane and he showed that these formulas are internally consistent. He observed that they can be obtained from the analogous formulas of spherical trigonometry by multiplying the sides of a triangle by an imaginary unit. Thus, he termed his plane "imaginary." He also used its properties to compute definite integrals, some known and some previously unknown. His effort was received by his Russian colleagues with disbelief and ridicule. But Lobachevsky remained convinced of the value of his discovery and published further articles about it over the next 25 years.

Bolyai's remarkably similar results constituted an article called "Supplement Containing the Absolutely True Science of Space, Independent of the Truth or Falsity of Euclid's [Fifth Postulate] (That Can Never Be Decided A Priori)." It appeared in 1832 as an appendix to a book by his father, Farkas, which surveyed previous attempts to prove the parallel postulate. J. Bolyai termed "absolute" a plane that could be either Euclidean or the alternative one with respect to properties of parallel lines.

Gauss made many attempts to prove the Fifth Postulate early in his mathematical life and then became convinced that the alternative plane must exist. But he never published any of this work, most likely because he felt it would have a cool reception. He was delighted to read J. Bolyai's and Lobachevsky's articles, and he noted in private correspondence that they had replicated some of his own results. In a letter to Farkas Bolyai he wrote of Janos's paper,

> Indeed the whole contents of the work, the path taken by your son, the results to which he is led, coincide almost entirely with my own meditations, which have occupied my mind partly for the last thirty or thirty-five years. So I remained quite stupefied. So far as my own work is concerned, of which up till now I have put little on paper, my intention was not to let it be published during my lifetime. Indeed the majority of people have not clear ideas upon the questions of which we are speaking, and I have found very few people who could regard with any special interest what I communicated to them on this subject....I am very glad that it is just the son of my old friend who takes the precedence of me in such a remarkable manner. [R. Bonola, *Non-Euclidean Geometry*, Dover Publications, New York, 1955, p. 100]

Gauss was also interested in Lobachevsky's work and promoted it through private letters to other European mathematicians. The publication of one of these letters, a few years after Gauss's death, made an especially strong impression and generated a surge of interest in the new geometry. Several important papers about it were written in the late 1860s. In particular, the Italian Eugenio Beltrami published two articles in 1868 that contain the first concrete models of this alternative plane. One of them is the example studied in this book (see Example 4 of Chapter 1, and Chapter 11) and is known as the Beltrami-Klein model. That is, Beltrami found a subset of the Euclidean plane on which he defined lines, angles, distance, and angle measure in such a way that the properties of the alternative plane hold. So he proved that the "imaginary" plane of Lobachevsky is as consistent as the Euclidean plane.

Beltrami's results firmly established the existence of this non-Euclidean plane and finally settled the issue of the Fifth Postulate. It turns out that Beltrami's model was implicit in the more general geometric structures introduced in 1859 by the Englishman Arthur Cayley. But Cayley did not see the link between his work and the new planes of Lobachevsky and Bolyai. It remained for another eminent German mathematician, Felix Klein, to make this connection in 1871. Klein also coined the term "hyperbolic" for the new geometry. Other models for the hyperbolic plane were discovered by the French mathematician Henri Poincaré in the 1880s. A great deal of work has been done on this plane and its higher-dimensional analogs up to the present day, and many applications have been found throughout broad areas of mathematics and mathematical physics.

Why did over 2100 years elapse before the true status of the Fifth Postulate and the existence of the hyperbolic plane was revealed? Certainly the nature of the Euclidean plane as an abstraction from everyday human experience lends it an aura of uniqueness. It seemed imperative to almost everyone who considered the question that no other planar structure is possible. It is not surprising that some of the would-be provers of the Fifth Postulate resorted in the end to an appeal to real-world objects. Alexis Claude Clairaut in the eighteenth century justified the existence of a rectangle (as opposed to a Saccheri quadrilateral with acute angles) by citing the "form of houses, gardens, rooms, walls." The various mathematicians described previously did not suffer such a blatant lapse of methodology, but they all worked under an imposing psychological burden. In fact, the eighteenth-century German philosopher Immanuel Kant believed that Euclid's assumptions existed as part of the structure of the human mind and that no correct reasoning about geometry was even possible without them. Lobachevsky, Bolyai, Gauss, and Beltrami proved him wrong.

Why didn't the existence of spherical geometry provide a clue that other surfaces should be investigated as possible sources for an alternative plane? The study of the sphere, because of its applications to astronomy and the calendar, existed in ancient Babylon and Egypt. The earliest mathematical treatise that has come down to us is "On the Rotating Sphere" by Autolycus in the fourth century B.C. The geometry of the spherical surface was developed to a level comparable to that of the Euclidean plane by Menelaus of Alexandria in his work, "On the Sphere," written in the first century A.D. The results of our text that pertain to the sphere are all essentially to be found in Menelaus's book. Later contributions, especially to spherical trigonometry, were made by many others, until the great Swiss mathematician Leonhard Euler put the subject in its modern form in the eighteenth century. So the sphere was not unknown to those pursuing the Fifth Postulate. But lines on the sphere, the "great circles" that divide a sphere into two equal-sized hemispheres (see Chapter 1), are finite in extent, whereas lines in the plane, by a (tacit) assumption of Euclid, are infinite. So the sphere seemed irrelevant to the issue of parallel lines in the plane. And of course, there are no parallel lines on the sphere.

Consideration of standard problems of Euclidean geometry was never likely to lead to the discovery of a model of the hyperbolic plane. In every model, either distance or angle measure must be modified from Euclidean distance or measure by some strange sort of twist, as we will begin to see in Chapter 1. It would take the development of an extensive theory of geometric surfaces, a knowledge of this theory as possessed by Beltrami and Klein, and the motivation prompted by the message of Lobachevsky, Gauss, and Bolyai in order to find such a model.

So the hyperbolic plane took a long time to arrive. Once it did, the scientific world's perception of geometry, and indeed of all of mathematics, was fundamentally changed. No longer could mathematics be regarded as simply the investigation of unique structures that had a prior existence as ideal objects in some philosophical sense. Now, alternative assumptions could be made from which different yet equally valid consequences could be derived. Thus mathematics became a more arbitrary and at the same time a more powerful enterprise. The broader perspective has led to deeper insights.

The discovery of the hyperbolic plane generated new concerns about the foundations of Euclidean geometry. The intuition and visualization that for over 2000 years had quietly resided in Euclid's postulates and arguments became suspect. New axiomatic treatments were written, with the goal of bringing Euclid's tacit assumptions to the surface and making all strands of the arguments explicit. The first of these foundational works was completed in 1882 by the German mathematician Moritz Pasch. One of his contributions was to clarify the distinction between undefined and defined terms. He recognized that he must present some terms as undefined, or primitive, as otherwise either an infinite or a circular sequence of definitions must follow. He saw that Euclid's definition of a point as "that which has no part" (see Appendix I) was not really a definition in any rigorous sense, as "part" has no meaning. Presumably, Euclid's purpose here was to provide some intuition; he never used this definition, or several others, in any of his arguments. Pasch let the word "point" be an undefined term, and his postulates stated all that he allowed to be assumed about it. Our text also leaves "point" as undefined. However, the notion of "betweenness" among points was also an undefined notion for Pasch, whereas we define it in terms of distance (and collinearity) based on an assumed preexistence of the real number system (see Chapters 5 and 6).

The most influential axiomatic study of Euclidean geometry was published in 1899 by the famous German mathematician David Hilbert. His treatment presented 21 axioms or postulates (and for us these terms mean the same thing; namely, assumptions) and showed that the full theory of Euclidean plane and solid geometry, including the development of the real numbers, follows. The American mathematician George David Birkhoff, in an article he published in 1932, assumed the prior existence of the real numbers and formulated just four postulates from which he showed that the theory of the Euclidean plane can be derived. Two of these postulates are essentially the Ruler and Protractor Axioms that we discuss in Chapter 13. A number of authors have considered the simultaneous development of the Euclidean and hyperbolic planes. The first axiomatic setting of which we are aware for the simultaneous study of the Euclidean, hyperbolic, and spherical planes occurs in the 1969 book *College Geometry* by David C. Kay (Holt, Rinehart and Winston, New York). Much but by no means all of the axiom system developed in our text borrows from Kay's system.

Problem Set 0

The solutions to these problems will have to be informal arguments based on informal recollections of concepts and results from Euclidean geometry. These should include the area of a rectangle and a (right) triangle, congruence of triangles, the interior of an angle, the angle sum of (the three angles of) a triangle, and the property (tacit in Euclid) that any line in a plane separates the plane into disjoint halfplanes. But note that the Fifth Postulate (or Playfair's Postulate) is not to be used except where indicated. Most of these concepts will be introduced and carefully developed in the chapters to follow, where rigorous proofs then will be feasible and expected.

1. Figure 0.2 appears in the ancient Chinese text *Chou Pei Suan Ching* with $a = 3$ and $b = 4$ and presents an implicit proof of the Pythagorean Theorem. Make this proof explicit through the following steps:

   (a) Note that each of the four $a \times b$ rectangles in the picture is split into two congruent right triangles by a diagonal whose length is denoted d.

   (b) Show that the quadrilateral $PQRS$ is a square (that is, each vertex angle is a right angle). You may assume that the three angles of a triangle add to $180°$.

   (c) Use the decomposition of $PQRS$ into four triangles and a square to find the area of $PQRS$ in terms of a and b.

   (d) Conclude that $d^2 = a^2 + b^2$.

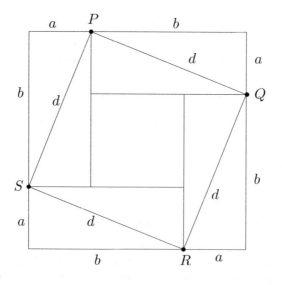

Figure 0.2

2. Euclid proved the Exterior Angle Inequality, which says that an exterior angle of a triangle is larger than either remote interior (that is, nonadjacent) angle, without using the Fifth Postulate. (See Chapters 1 and 15.) Use the Exterior Angle Inequality to

show that if line l crosses lines m and n so that the interior angles on one side add to two right angles (see Figure 0.3), then m and n are parallel. (Hint: Suppose that m and n meet and find a contradiction. Do *not* assume that the three angles of a triangle add to 180.)

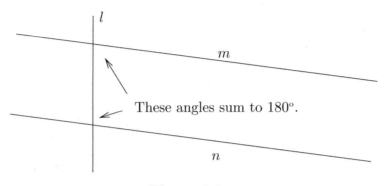

Figure 0.3

3. Show that Playfair's Postulate implies the Fifth Postulate. (Hint: Use Problem 2.)

4. Show that the Fifth Postulate implies Playfair's Postulate.

5. Show that if the Fifth Postulate holds, then the angle sum of any triangle equals two right angles (180°). (Hint: Consider the line through a vertex of the triangle that is parallel to the opposite side, as in Figure 0.4. This pair of parallel lines is crossed by each of the other two sidelines of the triangle. What can you say about the interior angles in these configurations?)

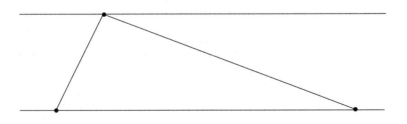

Figure 0.4

6. Assume a plane where Playfair's Postulate is false. That is, there are lines $\overleftrightarrow{QP}$ and $\overleftrightarrow{RP}$ both parallel to line l, as in Figure

0.5. Can you find more lines through P that are parallel to l? Explain.

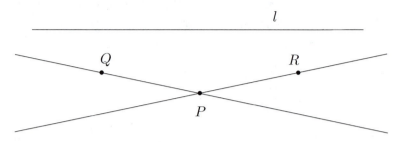

Figure 0.5

7. Consider the following claim (as in Figure 0.6):

(*) If X is any point in the interior of a proper angle $\angle QPR$, then there is a line through X that meets both sides of the angle.

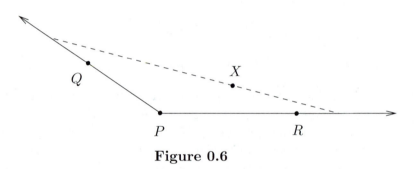

Figure 0.6

(a) Assume a plane where Playfair's Postulate is true. Show that (*) must also be true.

(b) Assume a plane where Playfair's Postulate is false. That is, there are lines $\overleftrightarrow{QP}$ and $\overleftrightarrow{RP}$ both parallel to line l, as in Figure 0.5. You may assume that l is contained in the interior of angle $\angle QPR$. Explain why (*) must be false for $\angle QPR$ and any point X on l. (That is, no line through X can meet both sides of $\angle QPR$.)

8. Suppose a plane where there is a (Saccheri) quadrilateral $ABCD$ with angles $\angle C$ and $\angle D$ right angles and $\angle A$ and $\angle B$ acute (as in Figure 0.7).

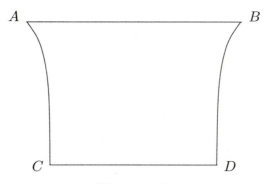

Figure 0.7

Show that the Fifth Postulate must then be false. (Hint: What does Problem 2 say about lines $\overleftrightarrow{AC}$ and $\overleftrightarrow{BD}$?)

9. Explain how a Saccheri quadrilateral can be obtained from a Lambert quadrilateral (with three right angles and an acute angle) by taking a mirror image across one side.

10. Assume that the Fifth Postulate is false; that is, angles $\angle SPQ$ and $\angle PQO$ sum to less than 180 in Figure 0.8, but lines $\overleftrightarrow{PS}$ and $\overleftrightarrow{QO}$ are parallel. Show that there exists a triangle with angle sum less than 180. (You may assume that points R exist on the ray $\overrightarrow{QO}$ with angle $\angle PRQ$ arbitrarily small; see Propositions 19.2 and 19.3.)

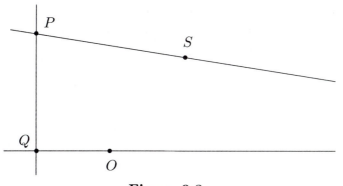

Figure 0.8

1 Five Examples

An important idea in this text is that there is more than one kind of geometry. The axiom system we will build describes not only the ordinary Euclidean plane but other interesting geometric structures as well. So the theorems we prove in the context of our axiom system yield information about a number of different examples all at once. The following five examples satisfy most of the axioms to be introduced, and three of them will satisfy all twenty-one of the axioms. It will be helpful to keep the examples in mind as we develop the general theory.

(1) $\mathbb{E}$: The usual *Euclidean plane*, points, and lines.

Coordinates: The points in $\mathbb{E}$ are in one-to-one correspondence with the ordered pairs of real numbers: Each point A corresponds to a pair of real numbers (x, y), called the *coordinates* of A, where the pair is assigned in the familiar way (Figure 1.1).

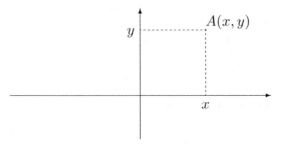

Figure 1.1

We often identify A with its pair of coordinates (x, y).

16

Equations of Lines. Each *nonvertical line* l in $\mathbb{E}$ consists of all points (x, y), where $y = mx + b$ for some fixed m and b. Each *vertical line* l consists of all (x, y), where $x = a$ for some fixed a.

For any two points $A(x_1, y_1)$ and $B(x_2, y_2)$, the *slope* of the line l through A and B is

$$m = \frac{y_2 - y_1}{x_2 - x_1} \qquad \text{(if } x_2 \neq x_1\text{)},$$

and an equation for l is given by

$$y - y_1 = m(x - x_1) \qquad \text{(if } x_2 \neq x_1\text{)}.$$

The *Euclidean distance* $e(AB)$ between A and B satisfies the formula

$$e(AB) = \sqrt{(x_2 - x_1)^2 + (y_2 - y_1)^2}.$$

Example. Let $A(1, 2)$, $B(-1, -3)$ be given. The line through A and B has slope $m = \frac{-3-2}{-1-1} = \frac{-5}{-2} = \frac{5}{2}$ and equation $y - 2 = \frac{5}{2}(x-1)$, or $y = \frac{5}{2}x - \frac{1}{2}$. Distance $e(AB) = \sqrt{(-3-2)^2 + (-1-1)^2} = \sqrt{25 + 4} = \sqrt{29}$.

Proposition 1.1. If $A(x_1, y_1)$ and $B(x_2, y_2)$ are on the line $y = mx + b$, then $e(AB) = |x_1 - x_2|\sqrt{m^2 + 1}$.

Proof. Problem 3.

(2) $\mathbb{M}$: The *Minkowski plane*, or *taxicab plane*.

$\mathbb{M}$ has the same points, lines, and coordinates as does $\mathbb{E}$, but *distance is different*: For any $A(x_1, y_1)$ and $B(x_2, y_2)$, define

$$d_{\mathbb{M}}(AB) = |x_2 - x_1| + |y_2 - y_1|.$$

So the Minkowski distance $d_{\mathbb{M}}(AB)$ is defined as the sum of the horizontal and vertical "ordinary distances" (Figure 1.2).

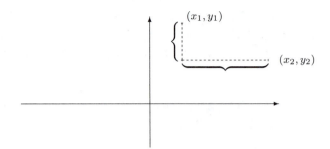

Figure 1.2

Example. Given $A(1,2)$, $B(-1,-3)$, then

$$d_{\mathbb{M}}(AB) = |-1-1| + |-3-2| = 2 + 5 = 7.$$

Proposition 1.2. If $A(x_1, y_1)$ and $B(x_2, y_2)$ are on the line $y = mx + b$, then $d_{\mathbb{M}}(AB) = |x_1 - x_2|(1 + |m|)$.

Proof. Problem 4.

(3) $\mathbb{S}(r)$: The (surface of the) *sphere* of radius r; that is, the *spherical plane*.

Once r is fixed, we shorten the notation to $\mathbb{S}$. We shall assume that our spheres are centered at the origin $(0,0,0)$ in three-dimensional space. Then $\mathbb{S}$ is the set of all (x, y, z) with $x^2 + y^2 + z^2 = r^2$. Points are as usual, and lines on $\mathbb{S}$ are defined to be the *great circles*. (A great circle is the intersection of the sphere with a plane that cuts the sphere in half; the equator is a great circle on the earth's surface.) Then any two points have a unique line joining them, *unless* they are opposite (antipodes), when they have infinitely many lines joining them (as infinitely many longitudinal great circles join the north and south poles).

Distance in $\mathbb{S}$: For points A, B on $\mathbb{S}$, define distance

$$d_{\mathbb{S}}(AB) = \quad \text{length of the minor (i.e., shorter) arc of the great circle (line) through } A \text{ and } B.$$

To compute $d_{\mathbb{S}}(AB)$ more easily, we first recall a formula for *arc length in a circle of radius r* (see Figure 1.3).

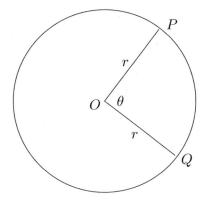

Figure 1.3

Let θ be the radian measure of $\angle POQ$. The angle that sweeps out the full circle has measure 2π, and the circumference is $2\pi r$. The sector formed by $\angle POQ$ makes up $\frac{\theta}{2\pi}$ of the full circle, so

$$\text{arc length } PQ = \frac{\theta}{2\pi} \cdot 2\pi r = \theta r.$$

For instance, $\theta = \frac{\pi}{3}$ implies that arc length equals $\frac{\pi}{3}r$.

Example. In $\mathbb{S}$ of radius 1, let $A = (1,0,0)$, $B = (\frac{\sqrt{3}}{2}, \frac{1}{2}, 0)$. We find $d_{\mathbb{S}}(AB)$: A, B lie on a great circle in the xy-plane (Figure 1.4).

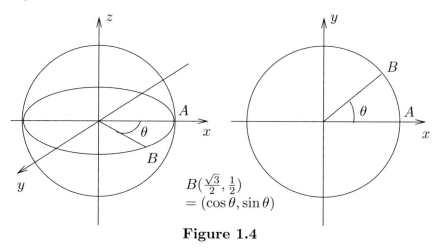

Figure 1.4

$\frac{\sqrt{3}}{2} = \cos\frac{\pi}{6}$ and $\frac{1}{2} = \sin\frac{\pi}{6}$, so $\theta = \frac{\pi}{6}$. So $d_{\mathbb{S}}(AB) = \theta r = \frac{\pi}{6}$.

An explicit formula for the spherical distance between two points, in terms of their coordinates, is given next. It follows from the distance formula for three-dimensional space and the Law of Cosines and is derived in Problem 10. Here, "$\cos^{-1}$" means the usual inverse cosine or arc cosine function, measured in radians.

If $P(a, b, c)$ and $Q(x, y, z)$ are points on the surface of the sphere of radius r centered at $(0, 0, 0)$ then

$$d_{\mathbb{S}}(PQ) = r \cos^{-1}\left(\frac{ax + by + cz}{r^2}\right).$$

Example. Let $A = (1, 0, 0)$ and $B = \left(\frac{\sqrt{3}}{2}, \frac{1}{2}, 0\right)$ on the sphere of radius 1, as in the previous example. Then by the formula,

$$d_{\mathbb{S}}(AB) = 1 \cdot \cos^{-1}\left(1 \cdot \frac{\sqrt{3}}{2} + 0 \cdot \frac{1}{2} + 0 \cdot 0\right) = \cos^{-1}\frac{\sqrt{3}}{2} = \frac{\pi}{6},$$

which agrees with the previous calculation.

Example. In $\mathbb{S}$ of radius 2, $P = \left(\frac{2}{3}, \frac{4}{3}, \frac{-4}{3}\right)$ and $Q = \left(\frac{6}{5}, 0, \frac{8}{5}\right)$. Then

$$d_{\mathbb{S}}(PQ) = 2 \cdot \cos^{-1}\left(\frac{\frac{2}{3} \cdot \frac{6}{5} + \frac{4}{3} \cdot 0 - \frac{4}{3} \cdot \frac{8}{5}}{2^2}\right)$$

$$= 2 \cdot \cos^{-1}\left(-\frac{1}{3}\right) = 3.82\ldots.$$

(4) $\mathbb{H}$: The *hyperbolic plane.*

$\mathbb{H}$ consists of all points *inside* (but not on) the unit circle in $\mathbb{E}$, that is, all (x, y) with $x^2 + y^2 < 1$.

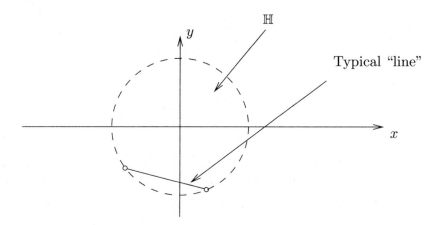

Figure 1.5

Lines in $\mathbb{H}$ are defined to be the chords of the circle (as in Figure 1.5). If A, B are two points in $\mathbb{H}$, define $d_{\mathbb{H}}(AB)$, the *distance* between them in $\mathbb{H}$, as follows: Draw the chord AB, and let M, N be the points where the chord meets the unit circle (M, N are in $\mathbb{E}$ but not in $\mathbb{H}$). Label so that B separates A and N (Figure 1.6).

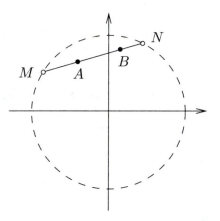

Figure 1.6

Let $e(PQ)$ denote the usual Euclidean distance between points, and *define*

$$d_{\mathbb{H}}(AB) = \ln\left(\frac{e(AN)e(BM)}{e(AM)e(BN)}\right),$$

where "ln" denotes the natural logarithm function. Since $e(AN) > e(BN)$ and $e(BM) > e(AM)$, we have $\frac{e(AN)}{e(BN)} > 1$ and $\frac{e(BM)}{e(AM)} > 1$.

Hence $\frac{e(AN)e(BM)}{e(AM)e(BN)} = \frac{e(AN)}{e(BN)}\frac{e(BM)}{e(AM)} > 1.$ It follows from a property of ln that $d_{\mathbb{H}}(AB) > 0$. Note that $d_{\mathbb{H}}(BA) = \ln\frac{e(BM)e(AN)}{e(BN)e(AM)} = d_{\mathbb{H}}(AB)$. Also,

$$d_{\mathbb{H}}(AB) = \left|\ln\left(\frac{e(AN)e(BM)}{e(AM)e(BN)}\right)\right| = \left|\ln\left(\frac{e(AM)e(BN)}{e(AN)e(BM)}\right)\right|.$$

So if absolute value is used in this way, then we need not worry about which point on the unit circle is marked M and which is marked N.

If $A = B$ in $\mathbb{H}$, take any chord through A and let M, N be as previously. Since $\frac{e(AN)e(AM)}{e(AM)e(AN)} = 1$, it is consistent with the preceding definition to set $d_{\mathbb{H}}(AA) = 0$.

Examples. If $A = (.8, 0)$, $B = (.9, 0)$, then $M = (-1, 0)$, $N = (1, 0)$ and

$$d_{\mathbb{H}}(AB) = \ln\left(\frac{(.2)(1.9)}{(1.8)(.1)}\right) = .7472144\ldots.$$

If $A = (0, 0)$, $B = (.999, 0)$, then $M = (-1, 0)$, $N = (1, 0)$ and

$$d_{\mathbb{H}}(AB) = \ln\left(\frac{(1)(1.999)}{(1)(.001)}\right) = 7.6004023\ldots.$$

If $A = (-\frac{1}{2}, 0)$, $B = (0, \frac{1}{2})$, then the line joining them is $y = x + \frac{1}{2}$, which meets $x^2 + y^2 = 1$ at

$$M = (-\frac{1}{4} - \frac{1}{4}\sqrt{7}, \frac{1}{4} - \frac{1}{4}\sqrt{7}), \quad N = (-\frac{1}{4} + \frac{1}{4}\sqrt{7}, \frac{1}{4} + \frac{1}{4}\sqrt{7}).$$

Then by Proposition 1.1, $e(AM) = |-\frac{1}{4} + \frac{1}{4}\sqrt{7}|\sqrt{2} = \frac{\sqrt{2}}{4}(\sqrt{7}-1) = e(BN)$, and $e(AN) = \frac{\sqrt{2}}{4}(\sqrt{7}+1) = e(BM)$. So

$$\begin{aligned} d_{\mathbb{H}}(AB) &= \ln\left(\frac{\frac{\sqrt{2}}{4}(\sqrt{7}+1)\frac{\sqrt{2}}{4}(\sqrt{7}+1)}{\frac{\sqrt{2}}{4}(\sqrt{7}-1)\frac{\sqrt{2}}{4}(\sqrt{7}-1)}\right) \\ &= \ln\frac{(\sqrt{7}+1)^2}{(\sqrt{7}-1)^2} = 2\ln\frac{(\sqrt{7}+1)}{(\sqrt{7}-1)} = 1.59\ldots. \end{aligned}$$

(5) $\mathbb{G}$: The *gap,* or *missing strip plane.*

The points of $\mathbb{G}$ are all those of $\mathbb{E}$ *except* those (x, y) with $0 < x \leq 1$ (Figure 1.7).

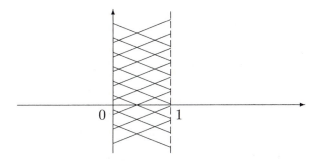

Figure 1.7

So the y-axis is part of $\mathbb{G}$, but the line $x = 1$ is *not* (and neither is any vertical line $x = a$ if $0 < a < 1$).

Lines in $\mathbb{G}$ are defined to be the same as in $\mathbb{E}$, *except* that for any nonvertical line $y = mx + b$, the part in the missing strip is deleted. So a typical nonvertical line l consists of all (x, y) with $y = mx + b$ (m, b fixed) *and* with $x \leq 0$ or $x > 1$ (Figure 1.8).

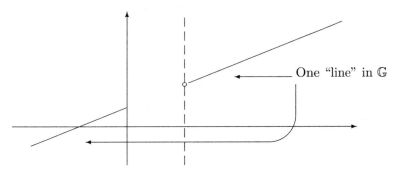

One "line" in $\mathbb{G}$

Figure 1.8

Distance: For points A, B in $\mathbb{G}$, we define $d_{\mathbb{G}}(AB)$ as follows. First, if A and B lie on opposite sides of the gap, let C be the point where segment $\overline{AB}$ meets the y-axis, and D the point where $\overline{AB}$ meets the vertical line $x = 1$ (D is not in $\mathbb{G}$; see Figure 1.9).

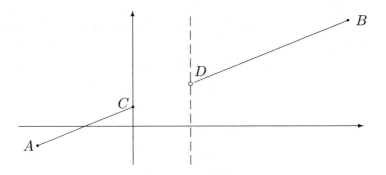

Figure 1.9

Now define

$$d_{\mathbb{G}}(AB) = \begin{cases} e(AB) & \text{for } A, B \text{ on the same side of the gap;} \\ e(AB) - e(CD) & \text{for } A, B \text{ on opposite sides.} \end{cases}$$

Example. Let $A = (-2, 0), B = (2, 0)$ (see Figure 1.10). Then $C = (0, 0), D = (1, 0)$ and

$$d_{\mathbb{G}}(AB) = e(AB) - e(CD) = 4 - 1 = 3.$$

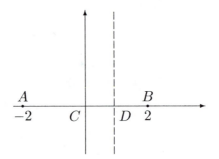

Figure 1.10

Example. $A = (-2, 0), B = (2, 1)$ lie on line l: $y = \dfrac{1}{4}x + \dfrac{1}{2}$, hence $C = (0, \dfrac{1}{2}), D = (1, \dfrac{3}{4})$ (see Figure 1.11).

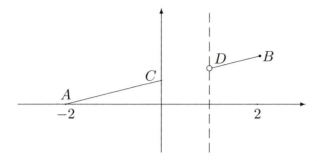

Figure 1.11

Now

$$e(AB) = \sqrt{(2+2)^2 + (1-0)^2} = \sqrt{17} \,,$$

and

$$e(CD) = \sqrt{(1-0)^2 + (\frac{3}{4} - \frac{1}{2})^2} = \sqrt{1 + \frac{1}{16}} = \sqrt{\frac{17}{16}} = \frac{1}{4}\sqrt{17} \,.$$

So $d_{\mathbb{G}}(AB) = \sqrt{17} - \frac{1}{4}\sqrt{17} = \frac{3}{4}\sqrt{17}.$

Each of the preceding five examples is a specific model of some geometric system. The similarities and differences among these models will help us to understand not only that alternative systems of geometry are just as real as is the Euclidean plane, but also that difficulties (and dangers) may arise when arguments use steps that are not explicitly justified.

For instance, consider the following proof of a form of the Exterior Angle Theorem in the Euclidean plane. This theorem plays a crucial role in the study of parallel lines (see, for example, the proof of Theorem 17.6). We will assume for the present discussion that you recall the relevant concepts from high school geometry. If not, don't worry. All of the ideas will be developed from scratch later in our more general setting.

Exterior Angle Inequality.   An exterior angle of a triangle is greater than either remote interior angle. That is, if $\triangle ABC$ is any triangle and D is on the extension of $\overline{BC}$ through C, then $\angle ACD$ is greater than each of $\angle A$ and $\angle B$ (Figure 1.12).

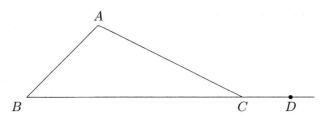

Figure 1.12

Proof.   Let M be the midpoint of $\overline{AC}$, and extend segment $\overline{BM}$ its own length through M to E.   Then lengths $AM = MC$ and $BM = ME$, while angle measures $\angle AMB = \angle CME$ (vertical angles are equal).   Thus $\triangle AMB \cong \triangle CME$ (by the Side-Angle-Side criterion for congruence of triangles).   So $\angle ECM = \angle BAM$ (corresponding angles of congruent triangles are congruent) (Figure 1.13).

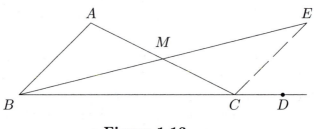

Figure 1.13

Since

$$
\begin{aligned}
\angle ACD &= \angle ACE + \angle ECD \\
&= \angle ECM + \angle ECD \\
&= \angle BAM + \angle ECD > \angle BAM = \angle A,
\end{aligned}
$$

we have that $\angle ACD > \angle A$.

To show that $\angle ACD > \angle B$, extend $\overline{AC}$ through C to F, forming $\angle BCF$ (Figure 1.14).   Then use the procedure of the first part of the proof to show $\angle BCF > \angle B$.   (Let N be the midpoint of $\overline{BC}$; extend $\overline{AN}$ its own length through N, etc.)

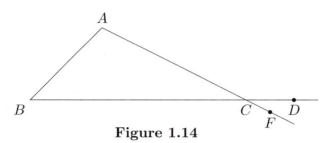

Figure 1.14

Finally, note that $\angle ACD = \angle BCF$ (vertical angles), and so $\angle ACD > \angle B$.

Now consider $\mathbb{S}$, on which "lines" are great circles. At a point where two great circles meet they form (four) angles, which may be given the "usual" angle measure. That is, assign the standard measure of the angles formed by the tangent lines to the great circles at the point of intersection. Assume that the notions of segment, triangle, and congruence all make sense in $\mathbb{S}$, in the most obvious way (they do!). Vertical angles have equal measure, and the Side-Angle-Side criterion for congruence of triangles is valid in $\mathbb{S}$. So it might seem plausible that the Exterior Angle Inequality holds here also. That is, a direct analog of the argument given for $\mathbb{E}$ might work in $\mathbb{S}$. However, consider the triangle $\triangle ABC$ shown in Figure 1.15.

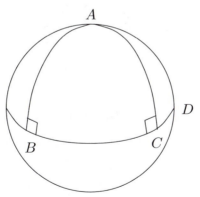

Figure 1.15

Think of A as the north pole, and B and C on the equator with $\angle A = \angle BAC > 90$. Then the exterior angle $\angle ACD = 90$, which is *less* than $\angle A$. So what goes wrong when we try to apply the preceding argument?

The problem lies in the assertion $\angle ACD = \angle ACE + \angle ECD$. There is a tacit assumption that the rays $\overrightarrow{CA}$, $\overrightarrow{CE}$, and $\overrightarrow{CD}$ are positioned so that the angle measures add in this way (i.e., that $\overrightarrow{CE}$ is "between" rays $\overrightarrow{CA}$ and $\overrightarrow{CD}$). There was no justification for this in the previous argument, except for "the picture" (Figure 1.13). When our rays, or segments, are parts of great circles in $\mathbb{S}$, a picture (see Problem 12) indicates that the relative positions may be quite different. The failure of the Exterior Angle Inequality in $\mathbb{S}$ is related to the fact that in the triangle $\triangle ABC$ under discussion, $\angle A + \angle B + \angle C$ *does not equal* 180. The extent to which the Exterior Angle Inequality does hold in $\mathbb{S}$ (and more generally) is discussed in Chapter 15.

It is important to note that assertions such as "ray $\overrightarrow{CE}$ is between $\overrightarrow{CA}$ and $\overrightarrow{CD}$" or "$\angle ACD = \angle ACE + \angle ECD$" *cannot be proved* from the given postulates of classical Euclidean geometry (see Appendix I). They follow only from unspoken assumptions about "betweenness" of rays and points, which are certainly plausible from pictures of configurations in the plane but which are never explicitly mentioned by Euclid. The preceding comparison shows that it pays to be aware of concepts such as "betweenness" and of the assumptions that we make about them. One of the goals of our rigorous development will be to make explicit all of the underlying suppositions about our geometry and thus to gain a deeper understanding of its fundamental concepts.

Problem Set 1

1. (a) In $\mathbb{E}$, find the distance between $A(-\frac{1}{3}, 0)$ and $B(0, \frac{1}{3})$.

   (b) In $\mathbb{M}$, find the distance between $A(-\frac{1}{3}, 0)$ and $B(0, \frac{1}{3})$.

   (c) In $\mathbb{H}$, find the distance between $A(-\frac{1}{3}, 0)$ and $B(0, \frac{1}{3})$.

   (d) In $\mathbb{S}$, (radius $r = 1$), find the distance between $C(0, 0, 1)$ and $D(0, -\frac{1}{\sqrt{2}}, \frac{1}{\sqrt{2}})$.

(e) In $\mathbb{S}$, (radius $r = \dfrac{1}{2}$), find the distance between

$$P\left(\frac{1}{4}, \frac{\sqrt{2}}{4}, -\frac{1}{4}\right) \text{ and } Q\left(\frac{1}{6}, -\frac{1}{3}, \frac{1}{3}\right).$$

(f) In $\mathbb{G}$, find the distance between $A(-2, -3)$ and $B(4, 6)$.

2. Find two points A, B in $\mathbb{H}$ such that $d_{\mathbb{H}}(AB) > 13$. Show the calculation that justifies your answer.

3. Prove Proposition 1.1.

4. Prove Proposition 1.2.

5. Let $A(x_1, y_1)$ and $B(x_2, y_2)$ be two points on opposite sides of the gap in $\mathbb{G}$ and on the line $l : y = mx + b$. *Derive a formula* for $d_{\mathbb{G}}(AB)$ in terms of x_1, x_2 and m.

6. Find (sketch) the unit circle in $\mathbb{M}$; that is, the set of all points that are (Minkowski) distance 1 from the origin $(0, 0)$.

7. Find all points of the form $P(x, 0)$ in $\mathbb{H}$ whose distance from $O(0, 0)$ is $\ln 2$. $\qquad(\mathbb{H})$

8. True or false: For all points, P, Q, R in $\mathbb{G}$,

$$d_{\mathbb{G}}(PQ) + d_{\mathbb{G}}(QR) \geq d_{\mathbb{G}}(PR)?$$

Justify your answer.

9. True or false: For all noncollinear points P, Q, R in $\mathbb{M}$,

$$d_{\mathbb{M}}(PQ) + d_{\mathbb{M}}(QR) > d_{\mathbb{M}}(PR)?$$

Justify your answer.

10. Let $P(a, b, c)$ and $Q(x, y, z)$ be points on the sphere $\mathbb{S}$ of radius r centered at $O(0, 0, 0)$, as in Figure 1.16.

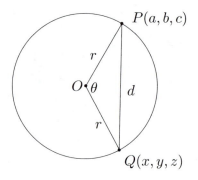

Figure 1.16

Let d be the Euclidean distance PQ and θ be the radian measure of $\angle POQ$.

(a) Recall the Law of Cosines for the triangle POQ and use it to show that $\cos\theta = (2r^2 - d^2)/2r^2$.

(b) Recall the Euclidean distance formula for points in three-dimensional space and use it and part (a) to show that
$$\cos\theta = \frac{ax + by + cz}{r^2}.$$

(c) Use (b) to derive that $d_{\mathbb{S}}(PQ) = r\cos^{-1}\left(\dfrac{ax + by + cz}{r^2}\right)$.

11. Let $P(1,0,0)$ be on the sphere $\mathbb{S}$ of radius 1. Find the set of points $Q(x,y,z)$ on $\mathbb{S}$ such that $d_{\mathbb{S}}(PQ) = 1$. That is, solve for the coordinates x, y, z and describe the set of points geometrically as well.

12. Draw a careful picture on $\mathbb{S}$ for the steps analogous to those in the argument for the Exterior Angle Inequality, but applied to $\triangle ABC$ on $\mathbb{S}$, as shown in Figure 1.15. Show the true relative positions of $\overrightarrow{CA}$, $\overrightarrow{CE}$, and $\overrightarrow{CD}$ (a ray here is half of a great circle). What can you conclude about $\angle ACD, \angle ACE$, and $\angle ECD$? (Use dotted lines to indicate parts of arcs on the "back side" of $\mathbb{S}$. Or, you can draw on the surface of a sphere (styrofoam ball, tennis ball, etc.).)

13. Let S be the part of the unit circle in $\mathbb{G}$ that lies to the right of the gap. That is, for $O = (0,0)$ and $P = (x,y)$,
$$S = \{P \in \mathbb{G} : d_{\mathbb{G}}(OP) = 1 \text{ and } x > 1\}.$$

(Set notation is reviewed in Chapter 4). Find an equation in x and y for the elements of S. (Hint: If $y = mx$, use Problem 5 of this problem set and note that $m = y/x$.)

14. Let C be the unit circle in $\mathbb{H}$. That is, for $O = (0,0)$ and $P = (x,y)$,
$$C = \{P \in \mathbb{H} : \ d_{\mathbb{H}}(OP) = 1\}.$$

Find an equation in x and y for the elements of C. In particular, show that C is a circle (of what radius?) in $\mathbb{E}$. (Hint: If $y = mx$, find the coordinates of M and N in terms of m, and then find $d_{\mathbb{H}}(OP)$ in terms of x and m. Note that $y = m/x$ if $x \neq 0$. Alternatively, we can set Euclidean distance $e(OP) = r$, derive a formula for $d_{\mathbb{H}}(OP)$ in terms of r, then set this expression equal to 1 and solve for r.)

15. Let D be the circle in $\mathbb{H}$ with radius 1 and center $Q = (\frac{1}{2}, 0)$. That is, for $Q = (\frac{1}{2}, 0)$ and $P = (x,y)$,

$$D = \{P \in \mathbb{H} : \ d_{\mathbb{H}}(QP) = 1\}.$$

(a) Find all points P on the x-axis that are also in D.

(b) Find all points P on the line $x = \frac{1}{2}$ that are also in D.

(c) Is D a circle in $\mathbb{E}$? Explain.

2 Some Logic

This chapter contains an informal presentation of logical concepts that will be used throughout. Our goal is simply to alert the reader to some common pitfalls. We focus initially on the meaning attached to words such as "all," "some," and "implies" when they appear in mathematical arguments; the misuse of these words often leads to trouble. Then we take a quick look at some ideas involved in the formulation of theorems and proofs, including the notions of "converse" and "contrapositive" and the method of proof by contradiction.

A statement is an assertion that is either true or false (and not both). We briefly discuss some words that are important in the logical analysis of statements. P and Q will denote arbitrary statements.

Equivalent: "P is equivalent to Q" means that P is true exactly when Q is true, and P is false exactly when Q is false. We denote this by "$P \Leftrightarrow Q$."

And: For "P and Q" to be true, *both* P and Q must be true.

Or: For "P or Q" to be true, either P is true or Q is true or both are true (inclusive or).

Sometimes when a statement of the form "P or Q" is used in ordinary speech, there is a tacit assumption that one of P or Q is necessarily false. There will be no such assumption attached to the word "or" in this text.

P and $Q \Leftrightarrow Q$ and P.　　P or $Q \Leftrightarrow Q$ or P.

All: A statement with the word "all" is an assertion about every member of a given collection of objects. Such a statement does not by itself require that any of the objects actually exists (even though such a prerequisite is sometimes implicit in everyday speech).

Example.   "All birds in the cage have feathers" is equivalent to "Whatever birds there are in the cage have feathers." An *all* statement is (vacuously) true if there are *none* of the things under discussion.

Example. Suppose that *there are no birds in the cage*. Then each of the following statements is *true*:

    All birds in the cage have four legs.

    All birds in the cage have three legs.

    All birds in the cage have red underwear.

Some: A statement with the word "some" is an assertion about one or more members of a given collection of objects. It does assert the existence of at least one member of the collection.

Example.   "Some birds in the cage have four legs" asserts that there is at least one bird that is in the cage and has four legs. So if we continue to assume that there are no birds in the cage, then the statement "Some birds in the cage have four legs" is *false*.

More Examples.

Some of the numbers $\{3, 5, 6\}$ are prime.	True
All of the numbers $\{3, 5, 6\}$ are prime.	False
Some of the numbers $\{3, 5, 7\}$ are prime.	True
All of the numbers $\{3, 5, 7\}$ are prime.	True
Some of the numbers $\{ \ \}$ are prime.	False
All of the numbers $\{ \ \}$ are prime.	True

Negation:

    "Not P" is true means P is false.
    "Not P" is false means P is true.

Not $(P$ and $Q)$ is equivalent to (Not P) *or* (Not Q).

This can be checked, as can the other claims of equivalence given in this section, by noting that the truth or falsity of the first statement (here, Not $(P$ and $Q)$) is always the same as the truth or falsity of the second ((Not P) or (Not Q) in this case) no matter what the truth or falsity of P, Q taken separately.

Example. "Not (The Bears are winning and the Broncos are winning)" is equivalent to "The Bears are not winning or the Broncos are not winning." In other words, "Not (The Bears and the Broncos are winning)" is equivalent to "Either the Bears or the Broncos are not winning."

Not $(P$ or $Q)$ is equivalent to (Not P) *and* (Not Q).

Example. "Not (The class is canceled or it is closed)" is equivalent to "The class is not canceled and it is not closed."

The negation of a statement that asserts that *all* members of a certain class have a certain property will say that *some* members of the class do *not* have the property.

Example. "Not (All birds have feathers)" is equivalent to

"Some birds do not have feathers."

We negate a statement that *some* members of a class have a certain property by saying that *all* members of the class do *not* have the property.

Example. "Not (Some students are lazy)" is equivalent to

"All students are not lazy."

The negation of a negation is equivalent to the original statement. That is,

$$\text{Not (Not } P) \Leftrightarrow P.$$

If/then: "If P then Q," equivalently written as "P implies Q" or "$P \Rightarrow Q$", means

Whenever P is true then so must Q be true.

$P \Rightarrow Q$ is (vacuously) *true* when P is *false*.

Example.

$$
\begin{array}{cc}
P & Q \\
\end{array}
$$

If $\underbrace{\text{circles have corners}}_{\text{False}}$ then $\underbrace{\text{I'm a monkey's uncle.}}_{\text{False}}$

$$\underbrace{\qquad\qquad\qquad\qquad\qquad\qquad}_{\text{True}}$$

So $P \Rightarrow Q$ is true whenever P is false (whether Q is true or false). But when P is true, then in order for $P \Rightarrow Q$ to be true, we must have that Q is true.

So $P \Rightarrow Q$ is equivalent to "Either P is false or Q is true." Hence

$P \Rightarrow Q$ is equivalent to   (Not P) or Q.

Negating $P \Rightarrow Q$:   From the preceding discussion,

Not $(P \Rightarrow Q)$	is equivalent to	Not $((\text{Not } P) \text{ or } Q)$
	which is equivalent to	Not (Not P) and Not Q
	which is equivalent to	P and Not Q.

Thus   Not $(P \Rightarrow Q)$ is equivalent to P and Not Q (an "and statement"). So to show $P \Rightarrow Q$ *false* (i.e., to show Not $(P \Rightarrow Q)$ *true*), we would have to find an example where P is true but Q is false.

Example.   To show "If a function is continuous then it is differentiable" is *false*, we need to find a function that is continuous but is *not* differentiable. The negation of

"If a function f is continuous, then f is differentiable"

is

"A function f is continuous and is not differentiable."

That is,

"There is a function f that is continuous and is not differentiable."

Example. The negation of

"If there is a major earthquake then the baseball game is canceled"

is

> "There is a major earthquake and the baseball game is not canceled."

In this and similar statements, the word "but" is considered grammatically equivalent to "and": "There was a major earthquake but the baseball game is not canceled."

Direct Proof:

The role of a proof is to establish that a conclusion follows logically from some given assumptions (hypotheses). The direct method of doing this involves presenting a series of statements, where each statement has an explicit reason which justifies its truth. Such a reason may be one or more of the hypotheses, or some previous assumption which remains in effect, or a definition, or a previously proved result, or a previous step in the proof itself. The final statement should be the desired conclusion. Many examples will follow shortly.

Proof by Contradiction:

To prove a statement by the method of contradiction, we

(1) assume the *negation* of the statement S we are trying to prove, that is, assume "Not S" is true, and then

(2) argue as usual, justifying each step, until we establish a statement that is the negation of a previously verified statement. This *contradiction* implies that "Not S" must be *false* and hence that S is true.

Sometimes the method of contradiction is used to establish one part of a larger argument but will not necessarily be in force throughout an entire proof. Some examples are given next and in the following chapters. When faced with devising a proof, we usually try the direct method first.

Never assume true that which you are trying to prove. Such procedure is always invalid. It is called "begging the question."

Example. The rooms of the Hotel Infinity all lie in a row and are numbered by the positive integers (Figure 2.1).

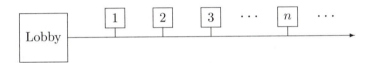

Figure 2.1

So there are infinitely many rooms. No, this is not a real-world situation, but it does provide a good context in which to illustrate some principles about proof. The following proofs are kept very simple, in order to let the methodology stand out clearly. In the discussion, a "guest" means someone who is registered for some room in the hotel and an "occupied room" is a room that has at least one guest registered for it.

Hypothesis A. For each positive integer n, there is an occupied room whose number is larger than n.

Theorem. Assume Hypothesis A. Then there are infinitely many guests in the hotel.

Proof. We use the method of contradiction. Suppose (toward a contradiction) that there are *not* infinitely many guests in the hotel. Then there is only a *finite* number (perhaps zero) of guests. If there are *no* guests, then Hypothesis A is immediately contradicted. So there must be a finite (positive) number of guests. Then the guests occupy only a finite (positive) number of rooms. We may compare the room numbers of the occupied rooms: Since there are only finitely many such numbers, there must be a largest one, say N.

Hence all rooms numbered above N are *vacant*. But Hypothesis A, applied to $n = N$, says that there is a room with number larger than N that is *occupied*. This *contradiction* means that our supposition "There are not infinitely many guests in the hotel" is *false*. Hence there must be infinitely many guests, and the theorem is proved.

Note: If we had started the preceding proof with "Suppose there are infinitely many guests in the hotel," we would have been begging the question. Our proof would have been invalid, no matter what we did next.

Question: Does it follow from Hypothesis A that every room is occupied?

Answer: No. Hypothesis A forces there to be an infinite set of rooms that are occupied, but there are many ways in which this can occur. For example, exactly those rooms with an even room number, or exactly those with a prime room number, or exactly those whose room numbers form some random increasing sequence may comprise the occupied rooms. It is important to be aware of exactly what a given assumption says and does not say.

We give two proofs for the next theorem: one that is direct and one by contradiction. Sometimes there is a choice, sometimes not.

Theorem: Assume Hypothesis A. Then there exist two guests whose rooms are more than 100 rooms apart.

First Proof. (Direct). Hypothesis A implies that there is a guest with room number greater than 1; say her room number is N. Then Hypothesis A, applied to $n = N + 100$, says that there is a guest with room number M such that $M > N + 100$. Then the two guests' rooms are more than 100 rooms apart, as $M - N > 100$.

Second Proof. Suppose (toward a contradiction) that there are *not* two guests whose rooms are more than 100 rooms apart. This means that *all* guests have rooms within 100 of one another. There *are* some guests, by Hypothesis A. Let one of the guests have room number, say, w. Then all guests have rooms numbered within 100 of w, and so have rooms numbered *at most* $w + 100$. But Hypothesis

A, applied to $n = w + 100$, says that there is an occupied room with number *larger than* $w + 100$. This is a contradiction.

Hypothesis B. The following statements are equivalent.

(i) Room 42 is occupied.

(ii) There is a guest over 7 feet tall in some room.

The assumption of Hypothesis B does *not* guarantee that either (i) or (ii) necessarily holds. It says only that *if* one of (i) or (ii) is true, then so is the other. The truth or falsity of (i) and (ii) will depend on other information, as in the following theorems. Notice that different assumptions lead to different results.

Theorem. Assume that Hypothesis B holds and that all the even-numbered rooms are occupied. Then there is a guest over 7 feet tall.

Proof. (By contradiction) Suppose that there is no guest over 7 feet tall. Then by Hypothesis B, Room 42 is *not* occupied. Since all even-numbered rooms are occupied, by assumption, it follows that 42 is not an even number. But $42 = 2 \cdot 21$, which *is* even, so we have a contradiction. Therefore, our supposition is false and so there *is* a guest over 7 feet tall.

A direct proof of the preceding theorem is easier and is worth writing out. See Problem 5 at the end of this chapter.

Theorem. Assume that Hypothesis B holds and that only the odd-numbered rooms are occupied. Then there is no guest over 7 feet tall.

Proof. Since 42 is not an odd number, it follows from our assumption that Room 42 is *not* occupied. Then Hypothesis B implies that the statement "There is a guest over 7 feet tall" must be *false*. So there is no guest over 7 feet tall.

We end this section by recalling the notions of "converse" and "contrapositive" of an if/then statement.

Converse: The converse of $P \Rightarrow Q$ is, by definition, $Q \Rightarrow P$.

When $P \Rightarrow Q$ is true, it is not necessarily the case that $Q \Rightarrow P$ is true. Sometimes it is, sometimes it isn't.

Examples.

> If f is a differentiable function, then f is a continuous function.
>   (True)
> If f is a continuous function, then f is a differentiable function.
>   (False)

(In the next pair of statements, x denotes a real number.)

> If $x = 0$, then $x^2 = 0$. (True)
> If $x^2 = 0$, then $x = 0$. (True)

The following two statements are converses of each other. It should be clear that they are *not* equivalent:

> If there is a major earthquake, then the baseball game is canceled.
> If the baseball game is canceled, then there is a major earthquake.

Contrapositive: The contrapositive of $P \Rightarrow Q$ is by definition

$$\text{Not } Q \Rightarrow \text{Not } P.$$

This is actually equivalent to $P \Rightarrow Q$, because

Not $Q \Rightarrow$ Not P is equivalent to	Not (Not Q) or Not P
which is equivalent to	Q or Not P
which is equivalent to	Not P or Q
which is equivalent to	$P \Rightarrow Q$.

Examples. If f is differentiable, then f is continuous. *Contrapositive*: If f is not continuous, then f is not differentiable.

If this course is easy, then it's a cold day in August. *Contrapositive*: If it's not a cold day in August, then this course is not easy.

If there was a major earthquake, then the baseball game is canceled. *Contrapositive*: If the baseball game is not canceled, then there was not a major earthquake.

Equivalence again. In order to prove a statement of the form "$P \Leftrightarrow Q$," two separate assertions must be verified:

$$P \Rightarrow Q \qquad (\text{If } P \text{ then } Q)$$

and its converse,

$$Q \Rightarrow P \qquad (\text{If } Q \text{ then } P).$$

"$P \Leftrightarrow Q$" that is, "P is equivalent to Q", is often stated as "P if and only if Q."

Problem Set 2

1. Form the *contrapositive* and *converse* of each of the following:

    (a) If a course is worthwhile, then it requires effort.

    (b) If Carl breaks the world record, then he wins a gold medal.

2. Form the *negation* of each of the following:

    (a) She sells sea shells and sheep bells.

    (b) If Sam finished his homework, then he enjoyed the weekend.

    (c) The heat wave breaks or we go swimming.

    (d) Every student has a point beyond which he cannot be forced to work.

    (e) All candidates are in debt or some voters are undecided.

    (f) If a man answers, then the caller hangs up.

    (g) Every quadrilateral has the sum of its interior angles equal to 360°.

    (h) Time and tide wait for no person.

3. Discuss the difference between the following statements:

All aspirin are not alike.
Not all aspirin are alike.

4. Assume Hypothesis A for the Hotel Infinity. Prove that there exist three guests, each of whom is at least 500 rooms away from each of the other two.

5. Write a direct proof of the theorem in the text that if Hypothesis B holds and all the even-numbered rooms are occupied, then there is a guest over 7 feet tall.

6. Assume that the center for the Chicago Bulls is 85 inches tall, that the Bulls are staying at the Hotel Infinity, and that Hypothesis B (as previously) holds. Use the method of contradiction to prove that Room 42 is occupied.

7. Assume Hypothesis B for the Hotel Infinity. Assume also that no room may contain more than four guests and that the combined height of the occupants of Room 3,749,861 is 29 feet. Prove that Room 42 is occupied.

8. Assume that you are the desk clerk at the Hotel Infinity on a night when each room has exactly one occupant. Assume that you may move guests from one room to another, but you may not evict anyone and you may not put more than one guest in a room.

   (a) Another person arrives who wants a room. Explain how you can accommodate him or her.

   (b) Now infinitely many more people (as many as there are positive integers) arrive and each wants a single room. Explain how you can accommodate them all.

3 Practice Proofs

The solutions of some logic puzzles are given in this chapter in the form of theorem and proof. Thus we illustrate some common proof techniques, in particular the elimination of possible cases by contradiction. All of these examples, except for Problems 6 and 7, are taken with only minor changes from C. R. Wylie, *101 Puzzles in Thought and Logic*, Dover Publications, New York, 1957 (pp. 2, 5, and Numbers 1, 2, 6, 57).

Example. Boronyenka, Pavlova, Revitskova, and Sukarek are four talented creative artists, one a dancer, one a painter, one a singer, and one a writer (though not necessarily respectively).

(1) Boronyenka and Revitskova were in the audience the night the singer made her debut on the concert stage.

(2) Both Pavlova and the writer have sat for portraits by the painter.

(3) The writer, whose biography of Sukarek was a best-seller, is planning to write a biography of Boronyenka.

(4) Boronyenka has never heard of Revitskova.

What is each woman's artistic field?

Theorem. Given the preceding information, Boronyenka is the dancer, Pavlova is the singer, Revitskova is the writer, and Sukarek is the painter.

Proof. Neither Sukarek nor Boronyenka is the writer, by (3). Also, Pavlova is not the writer, by (2). Thus Revitskova is the writer.

Boronyenka is not the singer, since (1) says she was in the singer's audience. So Boronyenka is either the painter or the dancer. Suppose (toward a contradiction) that Boronyenka is the painter. Then she has painted (and hence knows) the writer, by (2). We have shown that Revitskova is the writer, so Boronyenka would know Revitskova. This contradicts (4). Hence Boronyenka is *not* the painter, and so she is the dancer.

By (2), Pavlova has sat for a portrait, so she is not the painter. Thus Pavlova is the singer, and by elimination, Sukarek is the painter.

Example. Shorty was found shot to death one morning, and the police had three red-hot suspects behind bars by nightfall. That evening the men were questioned and made the following statements.

Buck: (1) I didn't do it.
           (2) I never saw Joey before.
           (3) Sure, I knew Shorty.

Joey: (1) I didn't do it.
           (2) Buck and Tippy are both pals of mine.
           (3) Buck never killed anybody.

Tippy: (1) I didn't do it.
           (2) Buck lied when he said he'd never seen Joey before.
           (3) I don't know who did it.

If one and only one of each man's statements is false, and if one of the three men is actually guilty, who is the murderer?

Theorem. Given the preceding information, Joey is the murderer.

Proof. If Tippy did it, then both of his statements (1) and (3) are false, which contradicts the hypothesis that only one of each man's statements is false. So Tippy didn't do it.

If Buck did it, then his statement (1) is false, and so by the hypothesis his statement (2) is true: Buck never saw Joey before. This forces Joey's statement (2), "Buck and Tippy are both pals of mine" to be false. But Joey's statement (3), "Buck never killed anybody," is also false if Buck did it. Now two false statements by Joey contradicts the hypothesis. So Buck didn't do it. Since one of the three is guilty, it must be Joey.

Problem Set 3

1. Given the assumptions of the example about Shorty, prove that Tippy knew who did it.

2. In a certain bank the jobs of cashier, manager, and teller are held by Flaherty, Lopez, and Rashad, though not necessarily respectively.

   > The teller, who was an only child, earns the least.

   > Rashad, who married Flaherty's brother, earns more than the manager.

   Find what position each woman fills, state your solution as a theorem, and prove it.

3. Clark, Daw, and Fuller make their living as carpenter, painter, and plumber, though not necessarily respectively.

   > The painter recently tried to get the carpenter to do some work for him but was told that the carpenter was out doing some remodeling for the plumber.

   > The plumber makes more money than the painter.

   > Daw makes more money than Clark.

   > Fuller has never heard of Daw.

   What is each man's occupation? State your solution as a theorem and prove it.

4. Clark, Jones, Morgan, and Smith are four men whose occupations are butcher, druggist, grocer, and policeman, though not necessarily respectively.

   > Clark and Jones are neighbors and take turns driving each other to work.

Jones makes more money than Morgan.

Clark beats Smith regularly at bowling.

The butcher always walks to work.

The policeman does not live near the druggist.

The only time the grocer and policeman ever met was when the policeman arrested the grocer for speeding.

The policeman makes more money than each of the druggist and the grocer.

What is each man's occupation?   State your solution as a theorem and prove it.

5. Bill Bianchi was shot to death at close range on a lonely country road late one night. The police soon established that the murder was committed by one of four men, Al, Jack, Joe, and Tom, and that the gun that was used belonged to one of the four. Each man was questioned and made the following statements, two and only two of which are true in each case.

Al: I didn't do it.
    Tom did it.
    Sure I own a gun.
    Joe and I were playing poker when Bill was shot.

Jack: I didn't do it.
    Al did it.
    Joe and I were at the movies when Bill was shot.
    Bill was shot with Joe's gun.

Joe: I was asleep when Bill was shot.
    Al lied when he said Tom killed Bill.
    Jack is the only one of us who owns a gun.
    Tom and Bill were pals.

Tom: I never fired a gun in my life.
    I don't know who did it.
    Joe doesn't own a gun.
    I never saw Bill until they showed me the body.

Who killed Bianchi? State your solution as a theorem, and prove it.

6. Smith, Jones, Brown, and Flynn are the math teacher, football coach, principal, and drivers' ed instructor (not necessarily in that order) at Euclidean High School. Use the following six clues to find what position each person holds, state your solution as a theorem, and prove it. Also, determine whether or not a referendum really passed.

   (1) Smith drives Jones to work in Smith's new car.

   (2) Brown cheered his son, the Euclidean High quarterback, to victory in every game this year.

   (3) Neither the math teacher nor the principal has any children.

   (4) The math teacher always walks to school.

   (5) The following statements are equivalent:

       (a) The Euclidean School District passed a referendum.

       (b) All the High School administrators live in Elliptic Estates.

       (c) The drivers' ed teacher bought a new Cadillac with his raise.

   (6) The following statements are equivalent:

       (a) The Euclidean High football team beat their cross-town rival Hyperbolic High this season.

       (b) The coach was the only school employee to buy a new car this year.

7. Arlo, Bruno, Carla, and Darla are among the residents of the city of Pointsville. Assume the following facts:

   (1) Arlo, Bruno, and Carla live (one each) on Locust, Maple, and Nut Streets, but not necessarily in that order.

   (2) Each person on Locust St. lives closer to every person on Nut St. than he or she does to every person on Maple St.

   (3) Whatever the largest distance is between the residents of Pointsville, Darla lives exactly that far from each person on Maple St.

(4) Arlo and Bruno live one mile apart. Carla lives two miles from each of Bruno and Darla. Carla and Arlo live three miles apart.

Given these assumptions, determine on which street each of Arlo, Bruno, and Carla lives. State your answers as a theorem and prove it.

4   Set Terminology and Sets of Real Numbers

We review in this chapter some standard notation and terminology for sets of numbers and other objects. We also present the Least Upper Bound Property of the real numbers, which is a cornerstone of a great deal of higher mathematics. We will need it in Chapter 5 in order to define what we call the diameter of a plane.

A set S consists of objects, called elements. A typical set S could be a set of numbers, a set of points, a set of functions, a set of lines, a set of apples, and so on. We read the symbols "$s \in S$" as "s is an element of S." Braces or curly brackets { } are often used to describe a set:

$$S = \{s : \ s \text{ is an element of } S\} = \{s : s \in S\}.$$

Example. The set of all even integers

$$\begin{aligned} &= \{x : \ x \text{ is an integer and } x = 2n \text{ for some integer } n\} \\ &= \{x : \ x = 2n \text{ for some integer } n\} \\ &= \{\ldots, -6, -4, -2, 0, 2, 4, 6, \ldots\}. \end{aligned}$$

The *empty set* (no elements) is denoted by $\emptyset$.

Ordered pair of elements (s, t): We take this as an obvious notion. The important thing about ordered pairs is that $(s, t) \neq (t, s)$ if $s \neq t$, whereas the sets $\{s, t\}$ and $\{t, s\}$ are the same.

For any sets S and T, the *Cartesian product* $S \times T$ of S and T

$$= \{(s, t) : \ s \in S, t \in T\}.$$

Example. Let $\mathbb{R}$ denote the set of all real numbers. Then $\mathbb{R} \times \mathbb{R} = \{(r, s) : r, s \in \mathbb{R}\}$, which corresponds to the set of all points in the Euclidean plane $\mathbb{E}$.

Two sets are called *equal* if they both consist of exactly the same elements. Given sets S and T, we say that S is a *subset* of T, or equivalently that S is *contained in* T (written $S \subseteq T$) if every element of S is also an element of T. That is,

$$S \subseteq T \text{ means that } s \in S \Rightarrow s \in T.$$

So $S = T \Leftrightarrow S \subseteq T$ *and* $T \subseteq S$.

The *union* of sets S and T is $S \bigcup T = \{x : x \in S \text{ or } x \in T\}$. So $S \bigcup T$ contains both S and T.

The *intersection* of S and T is $S \bigcap T = \{x : x \in S \text{ and } x \in T\}$. So $S \bigcap T$ is contained in both S and T.

The *difference* of S and T is $S - T = \{x : x \in S \text{ and } x \notin T\}$.

Let $\mathbb{R}$ denote the real numbers, and $<$ the usual "less than" ordering of $\mathbb{R}$. Geometrically, $a < b$ means that a lies to the left of b on the number line (Figure 4.1).

$$a \qquad\qquad\qquad\qquad\qquad\qquad b$$

Figure 4.1

We assume the following properties, for all $a, b, c \in \mathbb{R}$:

Exactly one of $a < b$, $b < a$, $a = b$ holds. (Note that $a \leq b$ and $b \leq a$ imply that $a = b$.)

If $a < b$, then $a + c < b + c$.

If $a < b$ and $c > 0$, then $ac < bc$.

If $a < b$ and $c < 0$, then $ac > bc$.

Definition. Let S be a nonempty set of real numbers. Then $b \in \mathbb{R}$ is an *upper bound* (u.b.) for S if $x \leq b$ for all $x \in S$.

Example. $S = $ closed interval $[0, 1] = \{x \in \mathbb{R} : 0 \leq x \leq 1\}$ has upper bounds $1, \frac{3}{2}, 2, 10$. *Any number $b \geq 1$ is an upper bound.*

Example. $S = $ open interval $(0, 1) = \{x \in \mathbb{R} : 0 < x < 1\}$ has an upper bound $1, \frac{3}{2}, 2, 10$. *Any number $b \geq 1$ is an upper bound,* and none of these upper bounds is in S itself.

Examples. $S = \mathbb{R}^+ = (0, \infty) = \{x \in \mathbb{R} : x > 0\}$ (all positive reals) has *no upper bound.*

$S = \{1, 2, 3, 4, \ldots, n, \ldots\}$ (all positive integers) has *no upper bound.*

$S = $ open interval $(-\sqrt{2}, \sqrt{2}) = \{x : x \in \mathbb{R} \text{ and } x^2 < 2\}$ has an upper bound, namely *any real number $b \geq \sqrt{2}$.*

$S = \{1, \frac{1}{2}, \frac{1}{3}, \frac{1}{4}, \ldots, \frac{1}{n}, \ldots\}$ has an upper bound, namely *any real number $b \geq 1$.*

To say that b is *not* an upper bound for a set S means that

Not (for all $x \in S$, $x \leq b$)

$\Leftrightarrow$ For some $x \in S$, it is *not* true that $x \leq b$

$\Leftrightarrow$ For some $x \in S, x > b$.

Definition. c is a *least upper bound* (l.u.b.) of a set S of real numbers if c is an upper bound for S and $c \leq b$ for all upper bounds b of S.

Examples. $S = [0, 1] \Rightarrow$ l.u.b. $= 1$. Here, the l.u.b. of S is in S.
$S = (0, 1) \Rightarrow$ l.u.b. $= 1$. Here, the l.u.b. of S is *not* in S.
$S = (-\sqrt{2}, \sqrt{2}) = \{x : x \in \mathbb{R} \text{ and } x^2 < 2\} \Rightarrow$ l.u.b. $= \sqrt{2}$.

Note: If a set S has a l.u.b., then it has only one, because if c, d are both least upper bounds for a set S, then c a l.u.b. and d an upper bound imply that $c \leq d$, while d a l.u.b. and c an upper bound imply that $d \leq c$. So $c \leq d$ and $d \leq c$ yield $c = d$.

We assume the following important property (see Figure 4.2):

Least Upper Bound Property of $\mathbb{R}$: If S is a nonempty set of real numbers that has an upper bound in $\mathbb{R}$, then S has a l.u.b. in $\mathbb{R}$.

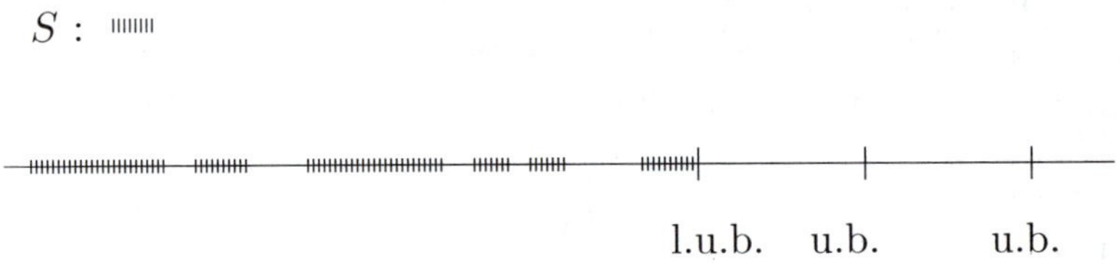

$$S: \quad \text{\tiny ||||||}$$

l.u.b.   u.b.   u.b.

Figure 4.2

Example. Let $\mathbb{Q}$ be the set of all rational numbers. That is,

$$\mathbb{Q} = \left\{ \frac{p}{q} : p, q \text{ integers and } q \neq 0 \right\},$$

so $\mathbb{Q} \subseteq \mathbb{R}$. Let $S = \{x : x \in \mathbb{Q} \text{ and } x^2 < 2\}$. Then S has rational upper bounds $2, \frac{3}{2}, \frac{23}{16}$. Any rational number $b > 0$ with $b^2 > 2$ is an upper bound for S. There is no smallest such rational number, so S has no *rational* l.u.b. The *real* l.u.b. of S is $\sqrt{2}$, which is *not* in $\mathbb{Q}$.

So the L.U.B. Property holds for $\mathbb{R}$ but *not* for $\mathbb{Q}$.

Example. Let $S = \{.3, .33, .333, .3333, \ldots\}$ so that S contains every *finite* decimal $.33\cdots3$. Then S has l.u.b. $\frac{1}{3}$, and $\frac{1}{3}$ is not in S.

Example. Let $S = \{3, 3.1, 3.14, 3.141, 3.1415, 3.14159, \ldots\}$ (continue with all the partial (finite) decimal expansions of π). Then S has l.u.b. π, and π is not in S.

Proposition 4.1. Let S be a nonempty set of real numbers that has a least upper bound b in $\mathbb{R}$. Let t be any real number such that $t < b$. Then there exists some $s \in S$ with $t < s \leq b$.

Proof. Problem 3.

Definition. Let S be a nonempty set of real numbers. Then $g \in \mathbb{R}$ is a *lower bound* for S if $g \leq x$ for all $x \in S$. Also, $h \in \mathbb{R}$ is a *greatest* *lower bound* (g.l.b.) for S if h is a lower bound for S and $h \geq g$ for all lower bounds g of S.

Example. Every negative real number is a lower bound for the interval $S = [0, 1]$, and the g.l.b for S is 0.

Proposition 4.2. Let S be a nonempty set of real numbers that has a lower bound in $\mathbb{R}$. Then S has a g.l.b. in $\mathbb{R}$.

Proof. Problem 4.

Problem Set 4

1. Given sets S and T, simplify each of the following and explain your answer.

   (a) $(S \cap T) \cup (S - T)$ (d) $(S - T) \cup (T - S)$

   (b) $(S \cap T) \cap (S - T)$ (e) $(S - T) \cap (T - S)$

   (c) $(S - T) \cup T$

2. For each set, either find the l.u.b. or explain why none exists:

   (a) $\{-2, -\frac{1}{2}, 0, \frac{4}{5}, \frac{3}{2}\}$

   (b) $\{x : x \in \mathbb{R} \text{ and } 5x^2 < 45\}$

   (c) $\{.6, .66, .666, .6666, \ldots\}$

   (d) $\{x^2 : x \in \mathbb{R} \text{ and } x < 2\}$

   (e) $\{x^3 : x \in \mathbb{R} \text{ and } x < 2\}$

   (f) $\{\frac{x}{3+x} : x \in \mathbb{R} \text{ and } x > 0\}$

   (g) $\{x : x = d_{\mathbb{S}}(PQ) \text{ for some points } P, Q \text{ in } \mathbb{S} \text{ (radius 1)}\}$

   (h) $\{x : x = d_{\mathbb{M}}(PQ) \text{ for some points } P, Q \text{ in } \mathbb{M}\}$

   (i) $\{x : x = d_{\mathbb{G}}(PQ) \text{ for some points } P, Q \text{ in } \mathbb{G}\}$

3. Prove Proposition 4.1.

4. Prove Proposition 4.2. (Hint: Let $T = \{-x : x \in S\}$ and show that T has a l.u.b.)

5. Assume Hypothesis A for the Hotel Infinity as presented in Chapter 2. Assume also that whenever x is a guest in Room m, y is a guest in Room n, and $m < n$, then y is taller than x. Let S be the set of all heights of all guests. Can S ever have an upper bound? Explain.

5 An Axiom System for Geometry: First Steps

In this chapter we begin to introduce the axioms for the system called an absolute plane, which includes as special cases not only the Euclidean plane but other geometric structures as well. Several chapters will be needed to present all of the axioms. First we give some remarks on axiom systems in general.

An *axiom system* is a carefully organized (one hopes!) description of one or more mathematical structures. A given system may describe several (or many) specific examples (models) at the same time. An axiom system consists of undefined terms (or semi-defined terms), definitions, axioms (basic assumptions about the terms), and theorems or propositions (consequences *proved* from the axioms). In a proof, each step must be justified by one or more of the following reasons:

Hypothesis of the statement being proved

Axiom

Previously proved theorem or proposition

Previous step in the proof

Definition

Basic logic

Now we start to unveil our system of the absolute plane.

Undefined Terms:

$\mathbb{P}$: a set of elements, called *points*

$\mathbb{L}$: a collection of subsets of $\mathbb{P}$, called *lines*

A function $d :\ \mathbb{P} \times \mathbb{P} \rightarrow \mathbb{R}$, called a *distance function*

For a point P and a line m, "P lies on m" and "m goes through P" just mean that $P \in m$. Two or more points are called *collinear* if there is a line that contains them; two or more lines are *concurrent* if there is a point contained in all of them.

For each ordered pair of points (P, Q) in $\mathbb{P} \times \mathbb{P}$, we denote $d(P, Q)$, the real number that d assigns to (P, Q), more simply by $d(PQ)$ or just by PQ when it is clear which particular function d is. Call PQ "the distance from P to Q."

Note that we intentionally are *not* specific about what points, lines, and distance really are. They are introduced simply as terms prescribing roles (vague roles, at the moment) that may be filled by many different systems of objects. What we may assume about points, lines, and distance is exactly what is set forth in the axioms to follow, no more and no less.

Any specified example of a set of points $\mathbb{P}$, set of lines $\mathbb{L}$, and distance function d will be referred to as a plane and usually will be abbreviated simply as $\mathbb{P}$. Admittedly, some of our examples may not seem very planar to the intuition. Try to keep an open mind. Unless stated otherwise, we will assume at any stage that our given abstract $\mathbb{P}$ satisfies whichever axioms have been introduced thus far. If $\mathbb{P}$ satisfies all of the axioms (the list of 21 axioms is completed in Chapter 13), then $\mathbb{P}$ will be called an *absolute plane*.

Axioms of Distance: For all points P and Q,

D1. (Positivity) $PQ \geq 0$.

D2. (Definiteness) $PQ = 0$ if and only if $P = Q$.

D3. (Symmetry) $PQ = QP$.

Definition. $\mathbb{D} = \{PQ :\ P, Q \in \mathbb{P}\}$.

$\mathbb{D}$ is the set of all distances that occur between points of $\mathbb{P}$; $\mathbb{D}$ is the range of the distance function d.   By Axiom D1, $\mathbb{D}$ is a set of nonnegative real numbers.

Definition. If $\mathbb{D}$ is nonempty and has an upper bound in $\mathbb{R}$, let ω be the l.u.b. of $\mathbb{D}$ (which exists by the Least Upper Bound Property). If $\mathbb{D}$ has no upper bound in $\mathbb{R}$, let $\omega = \infty$. (∞ is a symbol, not a real number.   But it is useful to write $r < \infty$ for all $r \in \mathbb{R}$.) ω is called the *diameter* of a given plane $\mathbb{P}$.

If $\omega = \infty$, then there are arbitrarily large distances; that is, for each real $r > 0$, there exist P, Q in $\mathbb{P}$ with $PQ > r$. (The choice of P and Q will depend on r.) Of course, $PQ < \infty$ for each P, Q in $\mathbb{P}$.

If $\omega < \infty$, then $PQ \leq \omega$ for all P, Q in $\mathbb{P}$. Proposition 4.1 shows that for any number $c < \omega$, there exist P, Q in $\mathbb{P}$ with $c < PQ \leq \omega$. When $\omega < \infty$, there may or may not exist points P, Q in $\mathbb{P}$ with $PQ = \omega$.

Axioms of Incidence:

I1. There are at least two different lines.

I2. Each line contains at least two different points.

I3. Each two different points lie in *at least* one line.

I4. Each two different points P, Q, *with $PQ < \omega$*, lie in *at most* one line.

Axioms I3 and I4 together imply that each two different points P, Q with $PQ < \omega$ lie together in *exactly one* line. So when $\omega = \infty$, each pair $P \neq Q$ of points determines a unique line.

If $\omega < \infty$ and if points P, Q have $PQ = \omega$, then P and Q may or may not lie in more than one line. (This will be illustrated in the following examples.)

Whenever there is a *unique* line through points P and Q, we denote it by $\overleftrightarrow{PQ}$.

Examples. The first five examples are as in Chapter 1.

$\mathbb{E}$ is the Euclidean plane, with distance $e(AB)$. The seven axioms hold. $\mathbb{D} = [0, \infty)$ (all nonnegative numbers), so $\omega = \infty$.

$\mathbb{M}$ consists of the same points and lines as $\mathbb{E}$, but $d_{\mathbb{M}}(AB) = |x_1 - x_2| + |y_1 - y_2|$, where $A = (x_1, y_1)$, $B = (x_2, y_2)$. The seven axioms hold. $\mathbb{D} = [0, \infty)$, so $\omega = \infty$.

$\mathbb{S}(r)$ is the set of points on the sphere of radius r, where the lines are great circles. $d_{\mathbb{S}}(PQ)$ is defined as the length of the minor arc of a great circle through P and Q. The seven axioms hold. $\mathbb{D} = [0, \pi r]$, so $\omega = \pi r$. See Figure 5.1.

πr is attained as the distance between any pair of opposite points.

Figure 5.1

For any points A, B with $d_{\mathbb{S}}(AB) = \omega = \pi r$, A and B must be opposite each other and there are infinitely many lines through A and B.

$\mathbb{G}$ is $\mathbb{E}$ with the strip $\{(x, y) : 0 < x \leq 1\}$ removed. Recall that, for C, D as shown in Figure 5.2, $d_{\mathbb{G}}(PQ) = e(PQ) - e(CD)$.

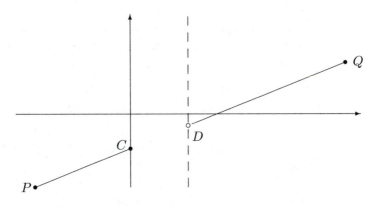

Figure 5.2

The seven axioms hold, $\mathbb{D} = [0, \infty)$, and $\omega = \infty$.

$\mathbb{H}$ consists of all points inside the unit circle in $\mathbb{E}$, where lines are defined as chords inside the circle. Recall that distance is defined as

$$d_{\mathbb{H}}(PQ) = \begin{cases} 0 & \text{if } P = Q \\ \ln\left(\dfrac{e(PN)e(QM)}{e(PM)e(QN)}\right) & \text{if } P \neq Q, \end{cases}$$

where M, N are points on the unit circle determined by P and Q, as shown in Figure 5.3.

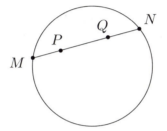

Figure 5.3

Clearly Axioms I1–I4 hold.   In fact, for each pair $P \neq Q$ of points, there is a unique line through P and Q.   The discussion in Chapter 1 shows that $\mathbb{H}$ satisfies Axioms D1–D3.   An exercise will show that $\mathbb{D} = [0, \infty)$ for $\mathbb{H}$, so $\omega = \infty$.

Definition.   A *model* for a given list of axioms is a specific mathematical structure whose existence is clear and for which each of the axioms on the list holds. (That is, each axiom is a true statement for the given structure.)

Each of $\mathbb{E}, \mathbb{M}, \mathbb{S}, \mathbb{G}$, and $\mathbb{H}$ is a model for the axioms D1–D3 and I1–I4. So is each of the next five examples.

Example 5.1. Let $\mathbb{P}$ be the set of all points inside the unit circle in $\mathbb{E}$, and $\mathbb{L}$ the set of all chords inside the circle (the same as for $\mathbb{H}$).   But for all $P, Q \in \mathbb{P}$, define distance $PQ = e(PQ)$, the usual Euclidean distance.   Then the seven axioms hold. $\mathbb{D} = [0, 2)$, so $\omega = 2$. But no distance $PQ = 2$ for any P, Q in $\mathbb{P}$. See Figure 5.4.

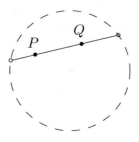

Figure 5.4

Example 5.2. Let $\mathbb{P}$ be the set of all points *inside or on* the unit circle in $\mathbb{E}$, and let $\mathbb{L}$ be the set of all chords. Again define distance $PQ = e(PQ)$. The seven axioms hold and $\mathbb{D} = [0, 2]$, so $\omega = 2$ here also. See Figure 5.5. Note that $AB = \omega$ exactly when A and B are opposite points on the unit circle. In this case, there is still a unique line through such A and B.

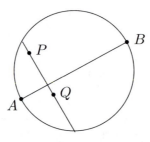

Figure 5.5

Example 5.3. Let $\mathbb{P} = \{x : x \in \mathbb{R} \text{ and } 0 < x < 1\}$. So here, "points" are numbers. Define a "line" to be any set of two numbers in $\mathbb{P}$. So each line in this example is a set of two points. Hence, given any $x_1 \neq x_2$ in $\mathbb{P}$, there is a unique line through x_1 and x_2, name $l = \{x_1, x_2\}$. (So Axioms I3 and I4 hold in this example, regardless of what ω turns out to be.) For each $x_1, x_2 \in \mathbb{P}$, we define *distance* by

$$d(x_1, x_2) = |x_1 - x_2|.$$

It is easy to see that this example satisfies the seven axioms and that $\mathbb{D} = [0, 1)$. So $\omega = 1$, which is *not* attained as a distance $d(x_1, x_2)$ for any x_1, x_2 in $\mathbb{P}$.

Example 5.4. Let $\mathbb{P} = \{A, B, C\}$ (so there are just three points in this model). *Define* the lines here to be the following three sets:

$$l = \{A, B\}, \qquad m = \{B, C\}, \qquad n = \{A, C\}.$$

We *define* all distances between points by the following table, where the distance XY between points X and Y is given by the entry in the row labeled by X, and the column by Y (so $AA = 0$, $AC = 17.5$, etc.):

	A	B	C
A	0	1	17.5
B	1	0	2
C	17.5	2	0

Then the seven axioms hold, $\mathbb{D} = \{0, 1, 2, 17.5\}$, and $\omega = 17.5$.

Example 5.5. (The Trivial Discrete Model) Let $\mathbb{P}$ be any set with at least three elements. Define a line to be any set of two elements in $\mathbb{P}$. So, as in Example 5.3, for any $X \neq Y$ in $\mathbb{P}$, there is a unique line through X and Y, namely $l = \{X, Y\}$. For each X, Y in $\mathbb{P}$, we define $d(XY) = 0$ if $X = Y$, and $d(XY) = 1$ if $X \neq Y$. It is easy to check (Problem 4) that the seven axioms hold, $\mathbb{D} = \{0, 1\}$, and $\omega = 1$.

The previous three examples show that while a model may satisfy the seven axioms, it does not have to resemble anything we are used to thinking of as geometry. In particular, the axioms so far guarantee nothing about the density, straightness, or connectedness of lines, whatever these terms may mean. Later axioms will place significant restrictions on the models we may construct.

Example 5.6. Let $\mathbb{P} = \{A, B\}$ (a set of two points). Define the lines to be $l = \{A, B\}$ and $m = \{A\}$. Define distances to be $AB = BA = 1$, $AA = BB = 0$. Then Axiom I2 is a false statement for this structure, while the other six axioms are true statements. So this example is *not* a model for the list of seven axioms.

We now start to derive consequences of the axioms and develop the general theory. All theorems about the abstract system are to be proved from *only* the axioms available.

Theorem 5.1. $\omega > 0$.

Proof. There *are* some lines, by Axiom I1. Each line contains at least two different points, by Axiom I2, so there are at least two points $P \neq Q$ in $\mathbb{P}$. Then $PQ > 0$ by Axioms D1 and D2. So $\mathbb{D}$ contains at least one positive number, namely PQ. Since ω is by definition an upper bound for $\mathbb{D}$, $\omega \geq PQ > 0$. So $\omega > 0$.

The next two results in this section are almost immediate consequences of Axioms I3 and I4. We include them to illustrate how statements can be rephrased through the use of basic logic.

Theorem 5.2. If m and n are different lines and distinct points A and B lie in both m and n, then $AB = \omega$.

Proof. Let A and B be distinct points in $\mathbb{P}$. Theorem 5.2 can be restated as follows: If A and B lie in more than one line, then $AB = \omega$.

Axiom I4 can be restated as follows: If $AB < \omega$, then A and B lie in at most one line. Since all distances are $\leq \omega$ by definition of ω, the statement "$AB < \omega$" is the negation of "$AB = \omega$." So Theorem 5.2 is the contrapositive of Axiom I4, and therefore it holds.

Theorem 5.3. If C and D are different points in $\overleftrightarrow{AB}$ and if $CD < \omega$, then $\overleftrightarrow{CD} = \overleftrightarrow{AB}$.

Proof. $CD < \omega$ implies that there is only one line through C and D, by Axioms I3 and I4. We call this line $\overleftrightarrow{CD}$. By hypothesis, $\overleftrightarrow{AB}$ goes through C and D. Therefore, $\overleftrightarrow{AB} = \overleftrightarrow{CD}$.

Problem Set 5

1. *Construct an example* of a set $\mathbb{P}$ (points), some subsets of $\mathbb{P}$ (lines), and a distance function $d : \mathbb{P} \times \mathbb{P} \to \mathbb{R}$, *such that* the diameter $\omega < \infty$ and all of the axioms except I3 are true statements for this example, *but* Axiom I3 is a false statement.

2. Same as #1, except replace "I3" with "I4."

3. Show that in the $\mathbb{H}$ model, $\mathbb{D} = [0, \infty)$. (Hints: Compute $d_\mathbb{H}(AB)$ (in terms of x) for $A = (0,0)$ and $B = (x,0)$, $0 < x < 1$. Then use the fact that $\ln$ sends the interval $(1, \infty)$ onto $(0, \infty)$.)

4. Verify that the Trivial Discrete Model satisfies the first seven axioms.

5. Show that the first seven axioms imply that there are at least three different points in $\mathbb{P}$.

6. Let $\mathbb{P} = \{1, 2, 3\}$, $\mathbb{L} = \{\{1\}, \{1, 2\}, \{2, 3\}\}$. Define distance by
$$d(P, Q) = P - Q$$
for all P, Q in $\mathbb{P}$ (equal or not). So, for example, $d(3, 2) = 3 - 2 = 1$.

   (a) Tell which of the seven axioms fail to hold in this example, and explain.

   (b) Find $\mathbb{D}$, the set of all distances (i.e., the image of d), and find w, the l.u.b. of $\mathbb{D}$.

7. Give an example of a plane (which satisfies the first seven axioms) in which all the points are collinear.

8. Let $\mathbb{P} = \{A, B, C, D\}$ (four distinct points). Let $\mathbb{L} = \{\{D\}, \{A, C\}, \{B, C\}, \{B, C, D\}\}$. Define distance by the following table:

	A	B	C	D
A	0	2	4	3
B	2	0	3	2
C	4	3	2	2
D	−3	2	2	0

   (a) Tell which of the seven axioms fail to hold in this example, and explain.

   (b) Find $\mathbb{D}$, the set of all distances, and find w, the l.u.b. of $\mathbb{D}$.

9. Suppose that $\mathbb{P}$ is a plane that satisfies the first seven axioms and in which there are four distinct points A, B, C, D (among others) and three distinct lines l, m, n (among others). Also assume the following:

(1) Each of l, m, n contains exactly one of A, B, C, and each of A, B, C is in exactly one of l, m, n (but not necessarily in that order);

(2) $XY > XZ$ for all $X \in l$, $Y \in m$, $Z \in n$;

(3) $YD = \omega$ (the diameter) for all $Y \in m$;

(4) $AB = 1$, $BC = 2 = DC$, $AC = 3$.

Explain the relationship of these hypotheses to those of Problem 7 of Chapter 2. Then find which of l, m, n contains which of A, B, C. State your answer as a theorem and prove it.

6 Betweenness, Segments, and Rays

In this chapter we introduce some important ideas about sets of collinear points. We investigate the familiar form that these concepts take in our five basic models ($\mathbb{E}, \mathbb{M}, \mathbb{S}, \mathbb{H}, \mathbb{G}$). We also note that in other examples that satisfy the first seven axioms, these notions may occur in strange ways.

The concept of betweenness for points on a line is rather subtle, especially in our abstract development. It is an idea that Euclid took for granted.

The next definition (and all others, unless noted otherwise) is stated in our general context, which currently means any set of elements $\mathbb{P}$ (points), certain subsets (lines), and a distance assigned to each pair of points so that the first seven axioms hold. Each time we introduce a new axiom, it is added to our abstract setup.

Definition. Point B lies *between* points A and C (written A-B-C) provided that

(a) A, B, C are different, collinear points, and

(b) $AB + BC = AC$.

This is our definition of betweenness, and all proofs regarding betweenness must be based on it. We need to see how this definition relates to our intuitive notion of betweenness in the various examples. First, we present a couple of general but elementary theorems.

Theorem 6.1. (Symmetry) A-B-C $\Leftrightarrow$ C-B-A.

65

Proof. Assume first that A-B-C. Now, C, B, and A are different collinear points since A, B and C are. Furthermore,

$$
\begin{aligned}
CB + BA &= BC + AB && \text{(by Axiom D3)} \\
&= AB + BC && \text{(by commutativity of } + \text{ in } \mathbb{R}) \\
&= AC && \text{(since } A\text{-}B\text{-}C\text{)} \\
&= CA && \text{(by Axiom D3).}
\end{aligned}
$$

Therefore, C-B-A by definition. We have proved that A-B-C $\Rightarrow$ C-B-A. Since this result holds for *any* three points that satisfy a betweenness relation (we have that X-Y-Z $\Rightarrow$ Z-Y-X; it's just a change of notation), it follows that C-B-A $\Rightarrow$ A-B-C as well.

Theorem 6.2. (Unique Middle) If A-B-C, then both B-A-C and A-C-B are false.

Proof. Problem 1.

Examples. $\mathbb{E}$: Let $A(x_1, y_1), B(x_2, y_2), C(x_3, y_3)$ be distinct points on a line $l : y = mx + b$.

Suppose that $x_1 < x_2 < x_3$, so that, intuitively, you cannot go from A to C along l without touching B. See Figure 6.1. Then, by Proposition 1.1,

$$
e(AB) = |x_1 - x_2|\sqrt{1 + m^2} = (x_2 - x_1)\sqrt{1 + m^2} \text{ (since } x_2 > x_1\text{).}
$$

Similarly, $e(BC) = (x_3 - x_2)\sqrt{1 + m^2}$ and $e(AC) = (x_3 - x_1)\sqrt{1 + m^2}$, and so

$$
\begin{aligned}
e(AB) + e(BC) &= (x_2 - x_1)\sqrt{1 + m^2} + (x_3 - x_2)\sqrt{1 + m^2} \\
&= (x_2 - x_1 + x_3 - x_2)\sqrt{1 + m^2} = (x_3 - x_1)\sqrt{1 + m^2} = e(AC).
\end{aligned}
$$

Thus A-B-C by the general definition.

Similarly, if $x_3 < x_2 < x_1$, then C-B-A and hence A-B-C by Theorem 6.1.

Conversely, suppose that A-B-C.

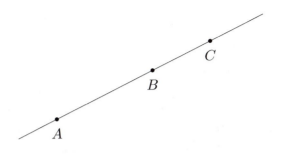

Figure 6.1

If $x_2 < x_1 < x_3$, then the preceding argument implies B-A-C. But both B-A-C and A-B-C true would contradict the Unique Middle Theorem (6.2). So it is *false* that $x_2 < x_1 < x_3$. Similarly, $x_3 < x_1 < x_2$, $x_1 < x_3 < x_2$, $x_2 < x_3 < x_1$ are all false. Hence, either $x_1 < x_2 < x_3$ or $x_3 < x_2 < x_1$.

Conclusion: In $\mathbb{E}$, A-B-C $\Leftrightarrow$ either $x_1 < x_2 < x_3$ or $x_3 < x_2 < x_1$. (It is easy to check that for points on a vertical line, a similar statement holds with respect to second coordinates.) In particular, any three distinct, collinear points in $\mathbb{E}$ satisfy some betweenness relation.

$\mathbb{M}$:   Let $A(x_1, y_1), B(x_2, y_2), C(x_3, y_3)$ be distinct points on a line $l : y = mx + b$. It follows from Proposition 1.2, in a manner analogous to the argument for $\mathbb{E}$, that

$$A\text{-}B\text{-}C \text{ (in } \mathbb{M}) \quad \Leftrightarrow \quad \text{either } x_1 < x_2 < x_3 \text{ or } x_3 < x_2 < x_1$$

and hence
$$A\text{-}B\text{-}C \text{ (in } \mathbb{E}) \Leftrightarrow A\text{-}B\text{-}C \text{ (in } \mathbb{M}).$$

$\mathbb{G}$:   It is an exercise (Problem 2) to show that for points A, B, C on a line l, all in $\mathbb{G}$ (as in Figure 6.2),

$$A\text{-}B\text{-}C \text{ in } \mathbb{E} \Leftrightarrow A\text{-}B\text{-}C \text{ in } \mathbb{G}.$$

$\mathbb{H}$:   It is a (longer) exercise to show that for points A, B, C on a line in $\mathbb{H}$ (that is, on a chord, which is part of a line in $\mathbb{E}$),

$$A\text{-}B\text{-}C \text{ in } \mathbb{E} \Leftrightarrow A\text{-}B\text{-}C \text{ in } \mathbb{H}.$$

(See Figure 6.3.)

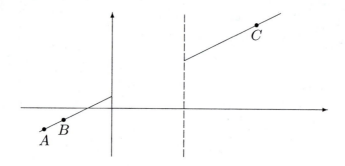

Figure 6.2

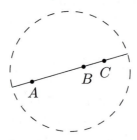

Figure 6.3

Note that for each of $\mathbb{M}$, $\mathbb{G}$ and $\mathbb{H}$, the case where A, B, C are on a vertical line must be checked separately.

$\mathbb{S}$ (radius r, so $\omega = \pi r$): Let A and C be two points on $\mathbb{S}$, not a pair of opposite points. Let l be the unique line (great circle) through A and C. We investigate which points B on l satisfy A-B-C. To do this, let A^* be the point opposite A and C^* the point opposite C (as in Figure 6.4).

Cases: (1) P on minor arc $\overparen{AC} \Rightarrow AP + PC = AC \Rightarrow A$-$P$-$C$

(2) P on minor arc $\overparen{A^*C}$ (including $P = A^*$) $\Rightarrow$
$$AC + CP = AP \Rightarrow A\text{-}C\text{-}P$$

(3) P on minor arc $\overparen{A^*C^*}$ (not including A^* or C^*) $\Rightarrow$
$$AC + CP > \omega > AP \Rightarrow \text{not } A\text{-}C\text{-}P$$
and $AP + PC > \omega > AC \Rightarrow$ not A-P-C
and $PA + AC > \omega > PC \Rightarrow$ not P-A-C

So in this case, there is *no* betweenness relation among $A, P,$ and C.

(4) P on minor arc $\overset{\frown}{C^*A}$ (including $P = C^*$) $\Rightarrow$
$$PA + AC = PC \Rightarrow P\text{-}A\text{-}C$$

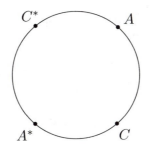

Figure 6.4

So given A, C on $\mathbb{S}$ with $AC < \omega$, we see that

A-B-$C \Leftrightarrow B$ is on the minor arc $\overset{\frown}{AC}$ of the great circle through A and C.

Given A, C with $AC = \omega$ (hence $C = A^*$), then A-B-C for *all* $B \neq A$, C on $\mathbb{S}$.

Example 6.1. (Inside-Out) Let $\mathbb{P} = \{A, B, C, D, E\}$. *Define* the lines here to be the following five sets:

$$l = \{A, B, C, D\}, \qquad m = \{A, E\}, \qquad n = \{B, E\},$$
$$u = \{C, E\}, \qquad\qquad v = \{D, E\}.$$

We *define* all distances between points as follows:

	A	B	C	D	E
A	0	3	1	2	4
B	3	0	2	1	4
C	1	2	0	3	4
D	2	1	3	0	4
E	4	4	4	4	0

Then the seven axioms hold, $\mathbb{D} = \{0, 1, 2, 3, 4\}$ and $\omega = 4$. It is left as an exercise (Problem 4) to verify that on line l, we have A-C-B, A-D-B, C-A-D and C-B-D. So each of C, D is between A and B, while at the same time each of A, B is between C and D – hence the name "Inside-Out."

The next definition puts the notion of "line segment" into our abstract context.

Definition. Let A, B be points with $0 < AB < \omega$. The *segment* $\overline{AB}$ is the set

$$\overline{AB} = \{A, B\} \cup \{X : A\text{-}X\text{-}B\}.$$

So the segment $\overline{AB}$ consists of all the points that are between A and B, along with A and B themselves.

We make the stipulation $AB < \omega$ in order that the following result will hold. It would be false (as a look at $\mathbb{S}$ shows) if we allowed "segments" $\overline{AB}$ with $AB = \omega$.

Proposition 6.3. (a) $\overline{AB}$ lies in exactly one line, namely $\overleftrightarrow{AB}$. (b) $\overline{AB} = \overline{BA}$. (c) If $X \in \overline{AB}$ with $X \neq B$, then $AX < AB$.

Proof. (a) Since $AB < \omega$, Axioms I3 and I4 imply that $\overleftrightarrow{AB}$ is the unique line through A and B. If $X \in \overline{AB}$ and $X \neq A$ or B, then the definition of $\overline{AB}$ yields $A\text{-}X\text{-}B$. So X is collinear with A and B, by definition of $A\text{-}X\text{-}B$. Then $X \in \overleftrightarrow{AB}$. The proofs of (b) and (c) are left as exercises (Problem 5).

Examples: Segments are now easy to picture in each of our five basic models (Figures 6.5 through 6.9).

Figure 6.5

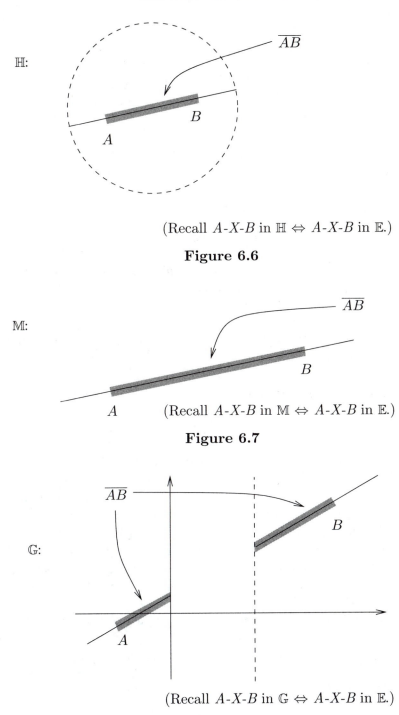

$\mathbb{H}$:

$\overline{AB}$

B

A

(Recall A-X-B in $\mathbb{H}$ $\Leftrightarrow$ A-X-B in $\mathbb{E}$.)

Figure 6.6

$\mathbb{M}$:

$\overline{AB}$

B

A (Recall A-X-B in $\mathbb{M}$ $\Leftrightarrow$ A-X-B in $\mathbb{E}$.)

Figure 6.7

$\overline{AB}$

B

$\mathbb{G}$:

A

(Recall A-X-B in $\mathbb{G}$ $\Leftrightarrow$ A-X-B in $\mathbb{E}$.)

Figure 6.8

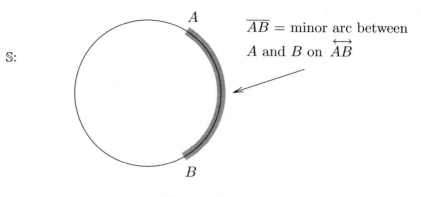

$\overline{AB}$ = minor arc between
A and B on $\overleftrightarrow{AB}$

$\mathbb{S}$:

Figure 6.9

Warning: "$\overline{AB}$" denotes a *segment*, that is, a particular *set of points*, while "AB" denotes the distance between A and B, a *number*. *Keep the two notations distinct!*

Now look at the Inside-Out example (Example 6.1) again. Line $l = \{A, B, C, D\}$ and A-C-B, A-D-B, C-A-D, and C-B-D. Since $AB = 3 = CD < 4 = \omega$, segments $\overline{AB}$ and $\overline{CD}$ are both defined. The definition of segment and the betweenness relations yield

$$\overline{AB} = \{A, B, C, D\} = \overline{CD} \qquad (= l).$$

So in this example, we have a segment with two *different* sets of "endpoints." We would like to rule out this sort of situation, but evidently we cannot do so on the basis of the axioms so far! More axioms will have to be adopted soon. It is interesting to observe, however, in the next proposition, that some general information about segments can be deduced just from our current, meager assumptions. Note that the proposition discusses a rather pathological situation: a given segment that has two possibly different sets of defining points.

Proposition 6.4.   Let A, B, C, D be collinear points with $0 < AB < \omega$, $0 < CD < \omega$, and $\overline{AB} = \overline{CD}$. Then

(a) either $\{A, B\} = \{C, D\}$ or $\{A, B\} \cap \{C, D\} = \emptyset$, and

(b) $AB = CD$.

Proof.   (a) Each of $\{A, B\}$ and $\{C, D\}$ is a set with two elements. There are three possibilities: $\{A, B\}$ and $\{C, D\}$ have two elements, one element, or no elements in common. If they have two elements in common, then $\{A, B\} = \{C, D\}$. If they have no common element, then $\{A, B\} \cap \{C, D\} = \emptyset$. So in either of these two cases, conclusion (a) holds. It remains to show that $\{A, B\}$ and $\{C, D\}$ cannot have exactly one common element.

Suppose (toward a contradiction) that $\{A, B\}$ and $\{C, D\}$ do have exactly one common element. This element is either $A = C$, or $A = D$, or $B = C$, or $B = D$. Now in the statement of this proposition, there is nothing assumed about B that is not also assumed about A. So we may switch their names (A becomes B, B becomes A) without any loss of generality. (This is called symmetry of hypothesis.) A similar remark holds for C and D. Thus we may assume that it is $A = C$ that comprises the intersection of $\{A, B\}$ and $\{C, D\}$. Then $A \neq D$, $B \neq D$, and $B \neq C$.

By definition of a segment, $D \in \overline{CD} = \overline{AB}$; and since $D \neq A$ or B, we must have A-D-B. Similarly, $B \in \overline{AB} = \overline{CD}$; and since $B \neq D$ or C, we have C-B-D. Then $A = C$ implies A-B-D. But A-D-B and A-B-D together contradict the Unique Middle Theorem. This contradiction means that $\{A, B\}$ and $\{C, D\}$ cannot have exactly one common element. The proof of (a) is complete.

(b) If $\{A, B\} = \{C, D\}$, then $AB = CD$, either by substitution or by Axiom D3. So by (a), we may assume that $\{A, B\} \cap \{C, D\} = \emptyset$. Thus $C, D \in \overline{CD} = \overline{AB}$ and $C \neq A$ or B, $D \neq A$ or B imply A-C-B and A-D-B. Similarly, $A, B \in \overline{AB} = \overline{CD}$ yields C-A-D and C-B-D, and hence $CA + AD = CD$ and $CB + BD = CD$. Adding these two equations and making suitable substitutions yields $2CD = 2AB$, hence $CD = AB$. The details are left as an exercise (Problem 6).

Definition. If $\overline{AB}$ is a segment, then the *length* of $\overline{AB}$ is the distance AB.

By Proposition 6.4(b), the length of a segment is uniquely determined, even though a segment may have more than one set of endpoints.

The critical geometric notion of "ray" can be presented in our abstract setup, as we proceed to do.

Definition. Let A, B be points with $0 < AB < \omega$. The *ray* $\overrightarrow{AB}$

is the set

$$\overrightarrow{AB} = \{A, B\} \cup \{X : \ A\text{-}X\text{-}B\} \cup \{X : \ A\text{-}B\text{-}X\}.$$

All points of $\overrightarrow{AB}$ are on the line $\overleftrightarrow{AB}$, which is called the *carrier* of $\overrightarrow{AB}$. A is called an *endpoint* of $\overrightarrow{AB}$.

So a ray is defined using two given points A and B (with $AB < \omega$) and consists of the segment $\overline{AB}$, together with all points X such that B is between A and X. Note that the roles of A and B in ray $\overrightarrow{AB}$ are *not* symmetric. A look at the following examples (and at many other examples) suggests that usually $\overrightarrow{BA} \neq \overrightarrow{AB}$.

Examples of Rays: The pictures in our five basic models are as shown in Figures 6.10 through 6.14.

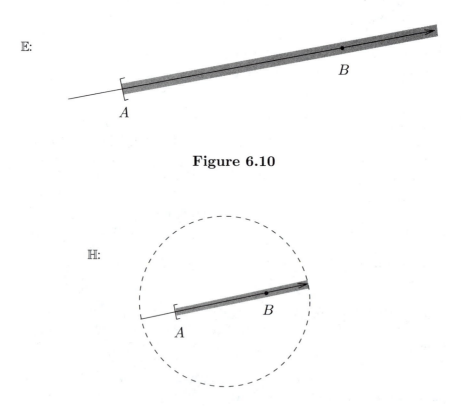

𝔼:

B

A

Figure 6.10

ℍ:

B

A

Figure 6.11

M:

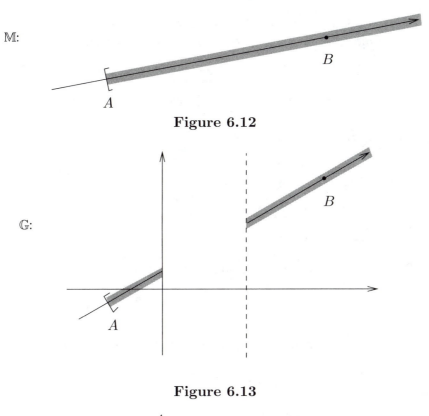

Figure 6.12

G:

Figure 6.13

S:

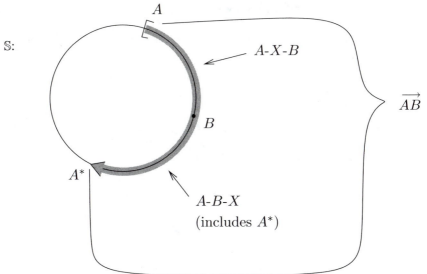

Figure 6.14

Note that in $\mathbb{S}$, $\overrightarrow{AB} = \overrightarrow{A^*B}$, so that a ray may have more than one endpoint. In fact, in the example Inside-Out we have

$$l = \{A, B, C, D\} \quad = \quad \overline{AB} = \overline{BA} = \overrightarrow{AB} = \overrightarrow{BA}$$
$$= \quad \overline{CD} = \overline{DC} = \overrightarrow{CD} = \overrightarrow{DC},$$

so that *every* point on this ray is an endpoint!

Problem Set 6

1. Prove Theorem 6.2.

2. Show that for any three points A, B, C on any line in $\mathbb{G}$,

$$A\text{-}B\text{-}C \text{ in } \mathbb{E} \Leftrightarrow A\text{-}B\text{-}C \text{ in } \mathbb{G}.$$

3. Show that for any three points A, B, C on any line in $\mathbb{H}$,

$$A\text{-}B\text{-}C \text{ in } \mathbb{E} \Leftrightarrow A\text{-}B\text{-}C \text{ in } \mathbb{H}.$$

You may proceed as follows:

(a) Suppose that $A\text{-}B\text{-}C$ in $\mathbb{E}$, which you may represent as shown in Figure 6.15.

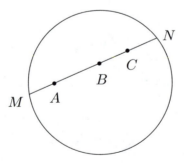

Figure 6.15

Use properties of ln to show that $A\text{-}B\text{-}C$ in $\mathbb{H}$.

    (b) For the converse, if A-B-C in $\mathbb{H}$, then A, B, C are collinear in $\mathbb{E}$; hence some betweenness relation among them holds in $\mathbb{E}$. Why must it be A-B-C?

4. Show that in Example 6.1, the relations A-C-B, A-D-B, C-A-D, and C-B-D all hold.

5. Prove Proposition 6.3, (b) and (c).

6. Complete the proof of Proposition 6.4(b).

7. Assume the first seven axioms. Suppose that A, B, X, Y are distinct, collinear points such that the distance between any two of them is less than ω and such that $Y \in \overline{AB}$, $X \in \overrightarrow{AB}$, $X \notin \overline{AB}$, and $B \in \overline{YX}$. Prove that $Y \in \overline{AX}$.

8. Construct an example of a plane $\mathbb{P}$ that satisfies the first seven axioms, with a segment $\overline{AB}$ and points X, Y in $\overline{AB}$ such that $XY > AB$.

9. Show that $\overrightarrow{AB} \cap \overrightarrow{BA} = \overline{AB}$.

10. Construct an example of a plane $\mathbb{P}$ that satisfies the first seven axioms, with a ray $\overrightarrow{AB}$ and points $X \neq Y$ in $\overrightarrow{AB}$ such that $AX = AY$.

11. Construct an example of a plane $\mathbb{P}$ that satisfies the first seven axioms, which is a finite subset of $\mathbb{S}$ and whose lines are all subsets of great circles in $\mathbb{S}$.

12. Assume the first seven axioms (and hence their consequences). Suppose that A, B, X, Y are distinct collinear points with $X \in \overline{AB}$, $A \in \overline{XY}$, and $XY = AB$. Prove that $AY = BX$.

7 Three Axioms for the Line

In this chapter we present three basic assumptions and some consequences for betweenness of points on a line. Some preliminary results and a definition appear first.

Proposition 7.1. If A-B-C and A-C-D, then A, B, C, D are distinct and collinear.

Proof. Problem 1.

Definition. A-B-C-D means that *all four* of the betweenness relations A-B-C, A-B-D, A-C-D, and B-C-D hold true.

Proposition 7.2. If A-B-C-D, then A, B, C, D are distinct and collinear, and D-C-B-A.

Proof. If A-B-C-D, then A-B-C, A-B-D, A-C-D, and B-C-D, by definition. Then A, B, C, D are distinct and collinear by Proposition 7.1. Theorem 6.1 implies that C-B-A, D-B-A, D-C-A, and D-C-B. It follows by the definition that D-C-B-A.

If A, B, C are collinear points such that the sum of each pair of distances between them (that is, $AB + BC$, $AC + CB$, $BA + AC$) is larger than ω, then no betweenness relation can hold among the three points. For if A-C-B, for example, then we would have $AC + CB = AB \le \omega$, which contradicts $AC + CB > \omega$. However, the converse statement, "If there is no betweenness relation among the distinct collinear points A, B, and C, then each of $AB + BC$, $AC + CB$, and $BA + AC$ is larger than ω," does *not* have to hold as a consequence of the first seven axioms (see Problem 3 at the end of this chapter). We *assume* the converse by the next axiom.

Betweenness of Points Axiom (Axiom BP):

If A, B, and C are different, collinear points and if $AB + BC \leq \omega$, then there exists a betweenness relation among A, B, and C.

Note that Axiom BP asserts that in fact, if any one of $AB + BC$, $AC + CB$, or $BA + AC$ is less than or equal to ω, then there is a betweenness relation among A, B, and C. So the contrapositive of Axiom BP may be stated as follows:

If there is no betweenness relation among the distinct collinear points A, B, and C, then each of $AB + BC$, $AC + CB$, and $BA + AC$ is larger than ω.

When $\omega = \infty$, then of course $AB + BC < \omega$ for any points A, B, C. So in this case, Axiom BP implies that *there exists a betweenness relation among any three distinct collinear points.* The discussion in Chapter 6 of betweenness in $\mathbb{E}$, $\mathbb{M}$, $\mathbb{G}$, and $\mathbb{H}$ shows that Axiom BP holds in each of these examples.

The analysis in Chapter 6 of betweenness in $\mathbb{S}(r)$ ($\omega = \pi r$) shows that a betweenness relation, say A-B-C, holds exactly when B is on the minor arc $\overarc{AC}$ of the great circle through A and C. So if there is *no* betweenness among collinear A, B, C in $\mathbb{S}$, then the three points do *not* all lie within one great semicircle, as in Figure 7.1.

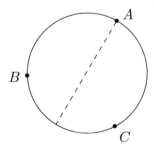

Figure 7.1

Hence, each of $AB + BC$, $AC + CB$, $BA + AC$ is larger than ω and Axiom BP holds.

Note that Axiom BP holds vacuously for any geometry where each line contains only two points. It also holds vacuously for the Inside-Out Model, since there is no set of three points on l that has no betweenness relation. So more assumptions will have to be made in order that segments and rays in our abstract geometry be as well behaved as we would like. This is done later. Meanwhile, some powerful results will be proved with only our current set of axioms.

Theorem 7.3. (Triangle Inequality for the Line) If A, B, and C are any three distinct collinear points, then $AB + BC \geq AC$.

Proof. Problem 2.

Theorem 7.4. (Rule of Insertion)

    (a) If A-B-C and A-X-B, then A-X-B-C.

    (b) If A-B-C and B-X-C, then A-B-X-C.

Proof. (a) By Proposition 7.1, A, X, B, C are distinct, collinear points. The definition of A-B-C and A-X-B yields

$$AC = AB + BC = AX + XB + BC. \qquad (*)$$

Now $XB + BC \geq XC$ by the Triangle Inequality. Hence, $(*)$ implies

$$AC \geq AX + XC.$$

But the Triangle Inequality also gives $AC \leq AX + XC$. Thus $AX + XC = AC$, and so A-X-C. Hence $XC = AC - AX$, which by $(*)$ equals $XB + BC$. Thus X-B-C, and so A-X-B-C by definition.

Part (b) follows from (a), Theorem 6.1, and Proposition 7.2. The details are left as an exercise (Problem 7).

The *definition* of A-B-C-D does not explicitly say that $AB + BC + CD = AD$. If A-B-C-D is true, then it is easy to show that $AB + BC + CD = AD$. On the other hand, if A, B, C, D are four distinct collinear points such that $AB + BC + CD = AD$, then it is possible (with some work and the use of the Triangle Inequality and the Rule of Insertion) to show that A-B-C-D holds. (See Problem 4.)

We need to introduce another axiom about betweenness of points. It is a little subtle, especially in light of the fact that when $w <$

∞, three given collinear points may satisfy no betweenness relation among themselves at all. The axiom says that *if* we are given three collinear points A, B, and C such that A-B-C, then any fourth collinear point must fit in among them in (at least) one of four precise ways.

Quadrichotomy Axiom for Points (Axiom QP):

If A, B, C, X are distinct, collinear points, and if A-B-C, then at least one of the following must hold:

$$X\text{-}A\text{-}B, \quad A\text{-}X\text{-}B, \quad B\text{-}X\text{-}C, \quad \text{or} \quad B\text{-}C\text{-}X.$$

The extent to which *exactly one* of the four conditions holds for a given point X is discussed in Theorem 9.2. Axiom QP is clearly true for $\mathbb{E}$, and hence for $\mathbb{M}, \mathbb{G}$, and $\mathbb{H}$, as follows from the discussion in Chapter 6.

Suppose that A-B-C in $\mathbb{S}$. Let B^* be the point opposite B (Figure 7.2).

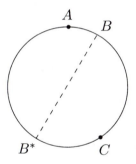

Figure 7.2

Let X be an point on the same line and distinct from each of A, B, and C. There are four possibilities for X:

$$X \text{ on minor arc } \overset{\frown}{B^*A} \Rightarrow X\text{-}A\text{-}B;$$
$$X \text{ on minor arc } \overset{\frown}{AB} \Rightarrow A\text{-}X\text{-}B;$$
$$X \text{ on minor arc } \overset{\frown}{BC} \Rightarrow B\text{-}X\text{-}C;$$
$$X \text{ on minor arc } \overset{\frown}{B^*C} \Rightarrow B\text{-}C\text{-}X.$$

Thus Axiom QP holds in $\mathbb{S}$. Note that if $X = B^*$, then *both* X-A-B and B-C-X are true.

Axiom QP holds vacuously in any plane in which there are at most three points on a line. It is *not* a consequence of Axiom BP (see Problem 5). But if a plane satisfies the first eight axioms (including BP), and *if the plane has infinite diameter* $(\omega = \infty)$, then QP follows as a theorem. (See Problem 11.) It is not hard to see that Axiom QP holds for the Inside-Out Model.

Note that when A-B-C is true, then $0 < AB < \omega$ and $0 < BC < \omega$, so line $\overleftrightarrow{AB}$ and rays $\overrightarrow{BA}$ and $\overrightarrow{BC}$ are defined. Axiom QP immediately implies that when A-B-C holds, then line $\overleftrightarrow{AB}$ is the union of the rays $\overrightarrow{BA}$ and $\overrightarrow{BC}$. This fact is critical for the notion of opposite rays, which is discussed in Chapter 9.

In the following result, we would like to force the stronger conclusion that when X, Y are two distinct points in a ray $\overrightarrow{AB}$ and each is different from A, then either A-X-Y or A-Y-X. But the Inside-Out plane, with $X = C$ and $Y = D$, shows that this conclusion is impossible without further assumptions. A sharper result will appear in the next chapter.

Proposition 7.5. If $X \neq Y$ are two distinct points on $\overrightarrow{AB}$, and both are different from A, then at least one of A-X-Y or A-Y-X or $X, Y \in \overline{AB}$ holds.

Proof. If $X = B$, then either A-X-Y or A-Y-X by definition of $\overrightarrow{AB}$. So we may assume that $X \neq B$ and similarly that $Y \neq B$. Now either A-X-B or A-B-X, and either A-Y-B or A-B-Y by definition of $\overrightarrow{AB}$.

If A-B-X and A-Y-B are both true, then the Rule of Insertion implies that A-Y-B-X. Hence A-Y-X, by definition of A-Y-B-X. So the conclusion holds in this case. Similarly, if A-B-Y and A-X-B, then A-X-Y. If A-X-B and A-Y-B, then the conclusion holds instantly.

So it only remains to consider the situation where A-B-X and A-B-Y. Now A-B-X and Axiom QP imply that one of Y-A-B, A-Y-B, B-Y-X, or B-X-Y holds. But Y-A-B or A-Y-B contradicts A-B-Y and the Unique Middle Theorem (6.2) and hence does not hold. If B-Y-X, then A-B-X and the Rule of Insertion yield A-B-Y-X, hence A-Y-X. If B-X-Y, then A-B-Y and the Rule of Insertion imply A-B-X-Y. So A-X-Y.

Results about rays are of little value unless we know that indeed there are some rays in the geometry. *Nothing in the axioms so far asserts that any rays exist.* We need two points A and B with $0 < AB < \omega$ in order to have a ray. But the Trivial Discrete Model (Example 5.5) satisfies all nine of the preceding axioms (vacuously in the case of BP and QP, as there are only two points on each line). Since $AB = 1 = \omega$ for all distinct points A and B, *there are no rays in the Trivial Discrete Model.* Similarly, there are no rays on any of the lines m, n, u or v of the Inside-Out plane.

So it is necessary to introduce a new axiom in order that our plane (each line, in fact) will have some rays. Note that this axiom is not needed in planes where $\omega = \infty$, since in that case it is already a true statement by Axioms I2, D1, and D2.

Nontriviality Axiom (Axiom N):

For any point A on a line l, there exists a point B on l with $0 < AB < \omega$.

Axiom N certainly holds in any example that satisfies the previous axioms *and in which $\omega = \infty$.* In particular, it is true for $\mathbb{E}$, $\mathbb{M}$, $\mathbb{H}$, and $\mathbb{G}$. It clearly holds in $\mathbb{S}$ also.

The Inside-Out and Trivial Discrete planes fail to satisfy Axiom N. But other finite models can be built to satisfy all ten axioms (see Problem 9). So another assumption will be necessary (and will be made in the next chapter) to ensure that our ultimate system of geometry has no finite examples.

The following consequence of Axiom N will be needed for our discussion of pencils of rays in Chapter 11.

Theorem 7.6. For any point A on a line l, there exists a point C *not* on line l with $0 < AC < \omega$.

Proof. Axiom I1 says that there is a line $m \neq l$. There is a point X on m by Axiom I2 and another point Y on m with $0 < XY < \omega$ by Axiom N. If X, Y are both on l, then $l = \overleftrightarrow{XY} = m$ by Axiom I4. This contradicts $m \neq l$.

So there exists a point, say X, on m with $X \notin l$. There is a line t through A and X (Axiom I3). Thus $t \neq l$. By Axiom N, there is a point C on t with $0 < AC < \omega$. If $C \in l$, then $l = \overleftrightarrow{AC} = t$ (Axiom I4), another contradiction. Hence $C \notin l$.

Problem Set 7

1. Prove Proposition 7.1. (Hints: A suitable definition and theorem (which?) show that A, B, C, D are distinct. Show that there is only one line through points A and C.)

2. Prove Theorem 7.3. (Hints: Consider separately the cases $AB + BC > \omega$ and $AB + BC \leq \omega$. Use Axiom BP.)

3. Construct an example of a plane that satisfies the first seven axioms and that has four collinear points A, B, C, D such that $AB + BC + CD = AD$ is true, but A-B-C-D is false. Does Axiom BP hold in your example? (Hint: Try to modify the Inside-Out Model.)

4. Let A, B, C, D be four distinct collinear points in a plane that satisfies the first eight axioms.

    (a) Prove that if A-B-C-D, then $AB + BC + CD = AD$.

    (b) Prove that if $AB + BC + CD = AD$, then A-B-C-D.

    (Hint: Use the Triangle Inequality and the Rule of Insertion.)

5. Construct an example of a plane that satisfies the first eight axioms but where Axiom QP fails to hold.

6. (a) Show that if $\omega = \infty$, then $\overrightarrow{AB} \cup \overrightarrow{BA} = \overleftrightarrow{AB}$.

    (b) Find an example in $\mathbb{S}$ to show that it is not always true that $\overrightarrow{AB} \cup \overrightarrow{BA} = \overleftrightarrow{AB}$.

7. Prove Theorem 7.4(b).

8. Assume that A, B, C, D are four distinct, collinear points with $B \in \overline{AC}$, $D \in \overrightarrow{AC}$, and $D \notin \overline{AC}$. Prove that $A \in \overrightarrow{DB}$.

9. Let $\mathbb{P} = \{A, B, C, D, E, F\}$ and $\mathbb{L} = \{\{A, B, C, D\}, \{A, E, F\},$ $\{B, E, F\}, \{C, E, F\}, \{D, E, F\}\}$. Define distance by the following table:

	A	B	C	D	E	F
A	0	3	1	2	4	2
B	3	0	2	1	4	2
C	1	2	0	3	4	1
D	2	1	3	0	4	1
E	4	4	4	4	0	5
F	2	2	1	1	5	0

Show that $\mathbb{P}$ satisfies the first 10 axioms. On which lines are there no betweenness relations?

10. Can we deduce from the first nine axioms that if $AC < \omega$ and A-B-C, then $\overline{AB} \cup \overline{BC} = \overline{AC}$? How about from the first 10 axioms? (Hint: Use the Inside-Out Model (Example 6.1) and/or Problem 9.)

11. Prove that if $\mathbb{P}$ satisfies the first eight axioms, and if $\omega = \infty$, then the statement of Axiom QP follows as a theorem. (So in the case of planes of infinite diameter, it is unnecessary to assume Axiom QP.) (Hint: Assume (toward a contradiction) that none of X-A-B, A-X-B, B-X-C, B-C-X holds. Then apply Axiom BP.)

12. Suppose that A, B, C are three distinct collinear points such that $AC \leq \frac{1}{2}AB$ and $BC \leq \frac{1}{2}AB$. Prove that A-C-B and $AC = BC = \frac{1}{2}AB$.

8   The Real Ray Axiom and Its Consequences

We now come to an assumption that forces lines in our abstract system to resemble more closely the lines in the five basic models. In particular, this axiom implies that each line contains infinitely many points.

Real Ray Axiom (Axiom RR):

For any ray $\overrightarrow{AB}$ and any real number s with $0 \leq s \leq \omega$ ($s < \omega$ when $\omega = \infty$), there is a point X in $\overrightarrow{AB}$ with $AX = s$.

Axiom RR says that the distances along a ray $\overrightarrow{AB}$, measured from point A, comprise all real numbers in the interval $[0, \infty)$ (if $\omega = \infty$) or $[0, \omega]$ (if $\omega < \infty$). Axiom RR may be used to exhibit lots of points on a ray. For example, to show that there exist three distinct points on $\overrightarrow{AB}$, we first take three distinct positive real numbers s_1, s_2, s_3. We choose them all less than ω if $\omega < \infty$ (plenty of such numbers exist). Axiom RR says that there exists a point X_1 on $\overrightarrow{AB}$ with $AX_1 = s_1$; that there exists a point X_2 on $\overrightarrow{AB}$ with $AX_2 = s_2$; and that there exists a point X_3 on $\overrightarrow{AB}$ with $AX_3 = s_3$. Since the distances s_1, s_2, s_3 from A are all different, the points X_1, X_2, X_3 must be different from one another. This produces the three distinct points.

There is nothing in Axiom RR that says explicitly that for a given s there is *only one* X in $\overrightarrow{AB}$ so that $AX = s$. For all we know at this stage of the game, there might be 19 different points X on some ray $\overrightarrow{AB}$ so that $AX = 68$. It turns out, however, that Axiom RR *plus* the other 10 axioms *imply* that there is only one such X. This uniqueness is proved in Theorem 8.6.

Axiom RR clearly holds for $\mathbb{E}$, $\mathbb{M}$, and $\mathbb{G}$. It is left as an exercise (Problem 7) to show that $\mathbb{H}$ satisfies Axiom RR. If $\overrightarrow{AB}$ is a ray in $\mathbb{S}$, then $\overrightarrow{AB}$ is a great semicircle (Figure 8.1).

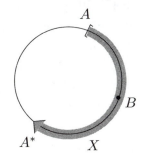

Figure 8.1

So all distances from 0 to ω appear in the form AX, for suitable $X \in \overrightarrow{AB}$. Thus Axiom RR holds in $\mathbb{S}$ also.

Theorem 8.1.   If $\omega = \infty$, then $\mathbb{D} = [0, \infty)$; if $\omega < \infty$, then $\mathbb{D} = [0, \omega]$.

Proof. Problem 1.

Theorem 8.2.   Each segment, ray, and line has infinitely many points.

Proof. Problem 2.

So no finite example, including Example 6.1, can possibly satisfy Axiom RR.

Next comes the promised sharpened version of Proposition 7.5.

Theorem 8.3. If $X \neq Y$ are points distinct from A on $\overrightarrow{AB}$, then one of A-X-Y or A-Y-X holds.

Proof. Suppose (toward a contradiction) that the conclusion is false. Then Proposition 7.5 implies A-X-B and A-Y-B.

Since $AB < \omega$, Axiom RR tells us that there exists a point E

on $\overrightarrow{AB}$ with $\omega \geq AE > AB$. Now E on $\overrightarrow{AB}$ means that either A-E-B, $E = B$, or A-B-E. But $AE > AB$ and Proposition 6.3(c) imply that A-E-B and $E = B$ are impossible. Hence A-B-E. Thus $EB < AE \leq \omega$.

Now A-X-B, A-B-E, and the Rule of Insertion imply A-X-B-E. Similarly, A-Y-B implies A-Y-B-E. Hence X-B-E and Y-B-E. So $X, Y \in \overrightarrow{EB}$ by Theorem 6.1, and $X, Y \notin \overline{EB}$ by the Unique Middle Theorem. Thus Proposition 7.5, applied to $\overrightarrow{EB}$, yields that either E-X-Y or E-Y-X. Hence, Y-X-E or X-Y-E (Theorem 6.1).

Suppose Y-X-E. We have A-Y-E (from A-Y-B-E), so the Rule of Insertion implies A-Y-X-E, hence A-Y-X. But this *is* the conclusion, which contradicts our previous supposition.

Hence X-Y-E. We have A-X-E (from A-X-B-E), so the Rule of Insertion yields A-X-Y-E, hence A-X-Y. This final contradiction proves the theorem.

The next theorem shows that there is nothing unique about the point B in the name for a ray $\overrightarrow{AB}$. Almost any other point on the ray may be used in its place.

Theorem 8.4. If C is any point on $\overrightarrow{AB}$ with $0 < AC < \omega$, then $\overrightarrow{AC} = \overrightarrow{AB}$.

Proof. We may assume $C \neq B$, hence either A-C-B or A-B-C. In either case, $B \in \overrightarrow{AC}$. If X is any point on $\overrightarrow{AC}$ that is different from each of A and B, then Theorem 8.3 and X, B on $\overrightarrow{AC}$ imply that either A-B-X or A-X-B. So $X \in \overrightarrow{AB}$. Thus $\overrightarrow{AC} \subseteq \overrightarrow{AB}$. Similarly, if $X \neq A$ or C is any point on $\overrightarrow{AB}$, then either A-C-X or A-X-C. Hence $\overrightarrow{AB} \subseteq \overrightarrow{AC}$.

We now rule out the possibility that a ray h can equal $\overrightarrow{AB} = \overrightarrow{CD} = \overrightarrow{EF}$ for *different* points A, C, E (i.e., that h has three or more endpoints).

Corollary 8.5. A ray has at most two endpoints.

Proof. Suppose that a ray h has three distinct endpoints A, C, and E. That is, $h = \overrightarrow{AB} = \overrightarrow{CD} = \overrightarrow{EF}$, where no two of points A, C, E

are equal. Theorem 8.3, applied to points A, C, E on $\overrightarrow{AB}$, yields that either A-C-E or A-E-C holds. The same theorem, applied to points A, C, E on $\overrightarrow{CD}$, also yields that either C-A-E or C-E-A is true. So by the Unique Middle Theorem, A-E-C must hold. But once again, Theorem 8.3, applied to A, C, E on $\overrightarrow{EF}$, implies that either E-A-C or E-C-A is true. This and A-E-C contradict the Unique Middle Theorem.

Theorem 8.6. (Unique Distances for Rays) For any ray $\overrightarrow{AB}$ and any real number s with $0 \le s \le \omega$, there is a *unique* point X on $\overrightarrow{AB}$ with $AX = s$. $X \in \overline{AB}$ if and only if $s \le AB$.

Proof. For $s = 0$, $X = A$ is the unique point in $\overrightarrow{AB}$ with $AX = 0$, by Axiom D2. If $X \ne Y$ are points distinct from A in $\overrightarrow{AB}$, then either A-X-Y or A-Y-X by Theorem 8.3. So either $AX < AY$ or $AY < AX$ by Proposition 6.3(c). This proves the first assertion. The second assertion follows from Proposition 6.3(c).

Proposition 8.7. Let $\overline{AB}$ be a segment and $X, Y \in \overline{AB}$. Then $XY \le AB$, and if $XY = AB$, then $\{X, Y\} = \{A, B\}$.

Proof. The conclusions hold if $\{X, Y\} = \{A, B\}$. So it suffices to assume that $Y \ne A$ or B and to prove that $XY < AB$. If $X = A$, then $XY = AY < AB$ by Proposition 6.3(c). If $X \ne A$, then either A-X-Y or A-Y-X, by Theorem 8.3. If A-X-Y, then $XY < AY < AB$, while if A-Y-X, then $XY < AX \le AB$, all by Proposition 6.3(c).

Proposition 8.8. If $\overline{AB} = \overline{CD}$, then $\{A, B\} = \{C, D\}$.

Proof. Proposition 6.4 implies $AB = CD$. Then $\{C, D\} = \{A, B\}$ by Proposition 8.7.

Definition. Given a segment $\overline{AB}$, A and B are called its *endpoints*. (Proposition 8.8 shows that a segment can have only one pair of endpoints; that is, there are two unambiguously determined endpoints.) The other points of $\overline{AB}$ are called *interior points* of $\overline{AB}$.

The *interior* of $\overline{AB}$, denoted by Int $\overline{AB}$ or $\overline{AB}^o$, means simply the set of interior points of $\overline{AB}$. Thus

$$\text{Int } \overline{AB} = \overline{AB}^o = \{X : A\text{-}X\text{-}B\}.$$

From Chapter 6, the *length* of $\overline{AB}$ is already defined as AB.

Proposition 8.9. In each segment $\overline{AB}$ there is a unique point M with the property that $AM = \frac{1}{2}AB$. Furthermore, $AM = MB$.

Proof. The first statement follows from Theorem 8.6 applied to ray $\overrightarrow{AB}$ and number $s = \frac{1}{2}AB$. Since M must satisfy A-M-B, we have

$$AB = AM + MB = \tfrac{1}{2}AB + MB.$$

Thus $MB = \frac{1}{2}AB = AM$.

Definition. The point M in Proposition 8.9 is called the *midpoint* of $\overline{AB}$. So it is characterized by the properties A-M-B and $AM = MB = \frac{1}{2}AB$.

Theorem 8.10. If U, V, Z are any three distinct points on a ray, then there is a betweenness relation among them.

Proof. Suppose first that (at least) one of the points, say U, is an endpoint. Then by Theorem 8.3, either U-V-Z or U-Z-V, which proves the assertion. So we may assume that *none* of U, V, Z is an endpoint. Let A be an endpoint of the ray, $A \neq U, V$ or Z.

The distances AU, AV, AZ are distinct, by Theorem 8.6. The hypotheses do not distinguish among U, V, Z, so we may choose notation to assume that $AU < AV < AZ$. Theorem 8.3 and $AU < AV$ imply that A-U-V. Similarly, A-V-Z holds. Then A-U-V-Z by the Rule of Insertion, and so U-V-Z.

The last result in this chapter is an easy consequence of the Real Ray Axiom applied to $\overrightarrow{BA}$, but it is convenient to have it stated explicitly. The proof is left as an exercise (Problem 5).

Proposition 8.11. Let A, B be any two points on line m with $0 < AB < \omega$. Then there exists a point C on m with C-A-B and $CB < \omega$.

Problem Set 8

1. Prove Theorem 8.1.

2. Prove Theorem 8.2.

3. Prove that if $0 < AB < \omega$, $X \neq Y$, A-X-B and A-Y-B, then either A-X-Y-B or A-Y-X-B. Does this same conclusion follow if $AB = \omega$?

4. Consider the following so-called "rule of two": If A, B, C, and D satisfy A-B-C and B-C-D, then A-B-C-D.

   (a) Show that this rule is valid if $\omega = \infty$.
   (b) Give an example to show that it is not always correct if $\omega < \infty$.

5. Prove Proposition 8.11.

6. (The "rule of two for rays") Prove that if A, B, C, D are points on a ray h (none necessarily an endpoint) so that A-B-C and B-C-D, then A-B-C-D.

7. Show via the following steps that $\mathbb{H}$ satisfies Axiom RR.

   Let $M = (r, mr + b)$ and $N = (t, mt + b)$ be the points of intersection of the line $y = mx + b$ with the unit circle, $r < t$. So $l = \{(x, mx + b) : r < x < t\}$ is a (typical nonvertical) line in $\mathbb{H}$. Let $A = (a, ma + b)$, $C = (c, mc + b)$ be two points on l. We will assume $a < c$, so that $r < a < c < t$ (as in Figure 8.2).

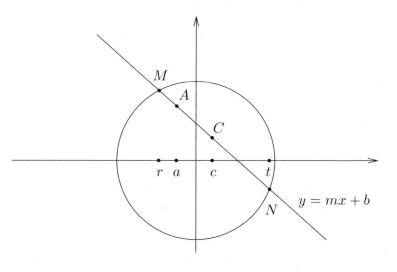

Figure 8.2

(a) Show that $X = (x, y)$ is on $\overrightarrow{AC}$ (in $\mathbb{H}$) if and only if $a \leq x < t$ and $y = mx + b$. (You may quote previous results, including problems.)

(b) For $X = (x, y) \neq A$ on $\overrightarrow{AC}$, show

$$AX = \ln \frac{(t - a)(x - r)}{(a - r)(t - x)}.$$

(c) For any real number $s > 0$, show that there exists an x with $a < x < t$ such that

$$\ln \frac{(t - a)(x - r)}{(a - r)(t - x)} = s,$$

and hence $AX = s$ for $X = (x, mx + b) \in \overrightarrow{AC}$.

8. Let $A = (0, 0)$ and $B = (.8, 0)$ in $\mathbb{H}$, and compute the midpoint of $\overline{AB}$. (Hint: It's not $(.4, 0)$.)

9. Let $A = (-1, 0)$ and $B = (3, 4)$ in $\mathbb{G}$, and compute the midpoint of $\overline{AB}$. (Hint: Write a typical point $X = (x, y)$ on $\overline{AB}$ in terms of x. Compute $d_{\mathbb{G}}(AX)$ and $d_{\mathbb{G}}(XB)$. You may assume that X is to the right of the gap.)

10. Let $A = (1, -1, 4)$ and $B = (3, 3, 0)$, two points on the sphere $\mathbb{S}$ of radius $r = 3\sqrt{2}$ and centered at the origin. Use the formula from Chapter 1 to compute $d_{\mathbb{S}}(AB)$. Then find the midpoint of $\overline{AB}$ on $\mathbb{S}$ by solving for the (unique, as it turns out) point $X = (x, y, z)$ on $\mathbb{S}$ such that $d_{\mathbb{S}}(AX) = \frac{1}{2}d_{\mathbb{S}}(AB)$ and $d_{\mathbb{S}}(BX) = \frac{1}{2}d_{\mathbb{S}}(AB)$.

11. Let $\overline{AB}$ be a segment in $\mathbb{E}$, hence also in $\mathbb{M}$. Prove that P is the midpoint of $\overline{AB}$ in $\mathbb{E}$ if and only if P is the midpoint of $\overline{AB}$ in $\mathbb{M}$. (Hint: If $\overleftrightarrow{AB}$ is a nonvertical line, apply Propositions 1.1 and 1.2.)

12. Prove that if $\omega < \infty$, then any ray is the union of two segments. (Hint: There is a point C on ray $\overrightarrow{AB}$ such that $AC = \omega$ (why?). Use Theorem 8.3 and the Rule of Insertion to show that $\overrightarrow{AB} = \overline{AB} \cup \overline{BC}$.)

13. Prove that if $\omega = \infty$, then no ray is the union of a finite number of segments.

9 Antipodes and Opposite Rays

We continue to show how abstract lines (given the first 11 axioms) must resemble those of $\mathbb{E}$ (when $\omega = \infty$) or of $\mathbb{S}$ (when $\omega < \infty$). In particular, when $\omega < \infty$ each point on a line has a unique opposite point or antipode on the line.

Theorem 9.1. Assume $\omega < \infty$. Let A be a point on line m. Then there exists a unique point A_m^* on m such that $AA_m^* = \omega$. Furthermore, if X is any other point on m, then $A\text{-}X\text{-}A_m^*$.

Proof. By Axiom N, there is a point X on m with $0 < AX < \omega$. For any such X, there exists a unique point A_m^* on $\overrightarrow{AX} \subseteq m$ with $AA_m^* = \omega$, by the Unique Distances Theorem (8.6). (At this stage, it may happen that A_m^* changes with X; we need to show that it does not.) Now $A\text{-}X\text{-}A_m^*$, again by Theorem 8.6 and the definition of $\overrightarrow{AX}$.

Suppose (toward a contradiction) that P is another point on m with $AP = \omega$. Then $P \notin \overrightarrow{AX}$. Quadrichotomy and $A\text{-}X\text{-}A_m^*$ imply that one of $P\text{-}A\text{-}X$, $A\text{-}P\text{-}X$, $X\text{-}P\text{-}A_m^*$, or $X\text{-}A_m^*\text{-}P$ holds. But $P\text{-}A\text{-}X$ would give $PX > PA = \omega$, and $A\text{-}P\text{-}X$ would yield $AX > AP = \omega$, each of which contradicts the definition of ω. If $X\text{-}P\text{-}A_m^*$, then $A\text{-}X\text{-}A_m^*$ and the Rule of Insertion force $A\text{-}X\text{-}P\text{-}A_m^*$, hence $A\text{-}X\text{-}P$. This says $P \in \overrightarrow{AX}$, another contradiction. Therefore, $X\text{-}A_m^*\text{-}P$. So $A_m^*P < PX \leq \omega$, and thus $\overline{A_m^*P}$ is defined. By the Unique Distances Theorem, there exists a point U with $A_m^*\text{-}U\text{-}P$, and hence $UP < A_m^*P < \omega$. Then $U \neq A$, since $AP = \omega$.

Now Axiom QP and A_m^*-U-P imply one of A-A_m^*-U, A_m^*-A-U, U-A-P, or U-P-A. But these imply, respectively, that $AU > AA_m^* = \omega$, $A_m^* U > A_m^* A = \omega$, $UP > AP = \omega$, or $UA > PA = \omega$, a contradiction in each case. So there is only *one* point A_m^* *on all of* m with $AA_m^* = \omega$. The first part of the argument now shows that for any X on m with $AX < \omega$, then A-X-A_m^*.

Definition. Assume $\omega < \infty$. Let A be a point on a line m. The unique point A_m^* on m with $AA_m^* = \omega$ (as in Theorem 9.1) is called the *antipode of A on m*.

Thus by Theorem 9.1, A and A_m^* are on m, $AA_m^* = \omega$, and A-X-A_m^* for all $X \neq A$, A_m^* on m.

In view of this definition, we shall call Theorem 9.1 the Antipode-on-Line Theorem.

It will be shown that, as a consequence of an axiom to be added later, for a given point A, the antipode A_m^* is the same point for *all* lines m through A (as is the case in $\mathbb{S}$). However, we can draw no such conclusion at this stage and must allow the possibility that A_m^* changes if m varies.

Theorem 9.2. (Almost-Uniqueness for Quadrichotomy) Suppose that A, B, C, X are distinct points on a line m and that A-B-C. Then *exactly one* of the following holds:

$$X\text{-}A\text{-}B, \qquad A\text{-}X\text{-}B, \qquad B\text{-}X\text{-}C, \qquad B\text{-}C\text{-}X,$$

with the solitary exception that both X-A-B and B-C-X are correct when $\omega < \infty$ and $X = B_m^*$.

Proof. Axiom QP says that *at least one* of X-A-B, A-X-B, B-X-C, B-C-X always holds.

Next, we show that no two of these hold simultaneously unless $X = B_m^*$ and precisely X-A-B and B-C-X are true (of the four relations). (Note that *if* $X = B_m^*$, then X-A-B and B-C-X do in fact hold by Theorem 9.1.)

Suppose that A-X-B. Since A-B-C is given, then A-X-B-C by the Rule of Insertion. Hence X-B-C holds, which means that both of B-X-C and B-C-X are false, by the Unique Middle Theorem (6.2).

Also, A-X-B implies that X-A-B is false, again by Theorem 6.2. So A-X-B implies that none of the other three relations holds.

Suppose that B-X-C. Since A-B-C, the Rule of Insertion implies that A-B-X-C. So A-B-X, which implies that X-A-B and A-X-B are false, by Theorem 6.2. Also, B-X-C means that B-C-X is false, so B-X-C implies that none of the other three relations holds.

So if more than one of X-A-B, A-X-B, B-X-C, or B-C-X hold, it must be that X-A-B and B-C-X hold, and A-X-B, B-X-C do not. So assume X-A-B and B-C-X.

Suppose (toward a contradiction) that $BX < \omega$. Then $\overrightarrow{BX}$ is defined, and X-A-B, B-C-X imply that $A, C \in \overrightarrow{BX}$. So one of B-A-C or B-C-A holds by Theorem 8.3. This contradicts Theorem 6.2 and the hypothesis A-B-C. Hence, $BX = \omega$ and $X = B_m^*$.

Corollary 8.5 showed that a ray has *at most* two endpoints. The next result proves that *when $\omega < \infty$, every ray has exactly two endpoints.* Specifically, whenever A is an endpoint of a ray with carrier m, then A_m^* is also an endpoint of the ray.

Proposition 9.3. Assume $\omega < \infty$. Let A, B be points on line m with $0 < AB < \omega$. Then

(a) $\overrightarrow{AB} = \overline{AB} \cup \overline{BA_m^*}$ and $\overline{AB}^o \cap \overline{BA_m^*}^o = \emptyset$.

(b) $\overrightarrow{AB} = \overrightarrow{A_m^* B}$, so that if A is an endpoint of a ray with carrier m, then so is A_m^*.

Proof. Problem 3.

The next theorem shows that the situation of Proposition 9.3 is the *only* way in which a ray can have two endpoints.

Theorem 9.4. If h is a ray with two endpoints A and B, then $\omega < \infty$ and $B = A_m^*$, where m is the carrier of h.

Proof. Suppose (toward a contradiction) that $AB < \omega$. Since B is an endpoint of h, Theorem 8.4 implies $h = \overrightarrow{BA}$. By Proposition 8.11 there is a point C on m with C-A-B and $CB < \omega$. Hence $C \in \overrightarrow{BA}$. So A, B, C are points of h, which has endpoint A. Then C-A-B

contradicts Theorem 8.3 and the Unique Middle Theorem. Thus $AB = \omega$, which implies that $\omega < \infty$ and $B = A_m^*$.

Definition. Let h be a ray. All points of h that are not endpoints of h are called *interior points* of h.

It follows from Theorem 9.4 that if $h = \overrightarrow{AB}$, then the interior points of h are exactly those points C on h with $0 < AC < \omega$.

Definition. Let $h = \overrightarrow{AB}$ be a ray. The *interior* of h is the set of all interior points of h and is denoted by h^o, $\overrightarrow{AB}^o$, or Int $\overrightarrow{AB}$. Thus

$$h^o = \overrightarrow{AB}^o = \text{Int } \overrightarrow{AB} = \{X \; : \; X \in \overrightarrow{AB} \text{ and } 0 < AX < \omega\}.$$

Note that $\overline{AB}^o \subsetneq \overrightarrow{AB}^o$. In particular, $B \in \overrightarrow{AB}^o$ while $B \notin \overline{AB}^o$.

Theorem 9.5. (Doubling a Segment) If segment $\overline{AB}$ has length $AB < \frac{\omega}{2}$, then there is an interior point C of $\overrightarrow{AB}$ so that $\overline{AC}$ has midpoint B.

Proof. Since $0 < AB < \frac{\omega}{2}$, we have $0 < AB < 2AB < \omega$. By Axiom RR, applied to ray $\overrightarrow{AB}$ and distance $s = 2AB$, there is a point C with A-B-C and $AC = 2AB < \omega$. So C is an interior point of $\overrightarrow{AB}$, and A-B-C says that $B \in \overline{AC}$. Since $AB = \frac{1}{2}AC$, B is the midpoint of $\overline{AC}$.

Definition. Two rays with the same endpoint whose union is a line are called *opposite* rays. (See Figure 9.1.)

$$A$$

Figure 9.1

Theorem 9.6. (Opposite Ray Theorem) If B-A-C, then $\overrightarrow{AB}$ and $\overrightarrow{AC}$ are opposite rays. Furthermore, for $m = \overleftrightarrow{AB}$,

$$\overrightarrow{AB} \cap \overrightarrow{AC} = \begin{cases} \{A\} & \text{if } \omega = \infty \\ \{A, A_m^*\} & \text{if } \omega < \infty. \end{cases}$$

Proof. Since $AB + AC = BA + AC = BC \leq \omega$ (by the definitions of ω and B-A-C), we have $AB < \omega$ and $AC < \omega$. So by Axiom I4, $\overleftrightarrow{AB}$ is the unique line through A and B, and $\overleftrightarrow{AC}$ is the only line through A and C. Now A, B, and C are collinear, since B-A-C, so $\overleftrightarrow{AB} = \overleftrightarrow{AC}$. Let $m = \overleftrightarrow{AB}$. Since $AB < \omega$ and $AC < \omega$, rays $\overrightarrow{AB}$ and $\overrightarrow{AC}$ are defined as subsets of m.

If $X \neq A$, B, or C is on m, then one of the following must hold, by Axiom QP:

$$X\text{-}B\text{-}A, \qquad B\text{-}X\text{-}A, \qquad A\text{-}X\text{-}C, \qquad \text{or} \qquad A\text{-}C\text{-}X. \qquad (*)$$

The first two cases of $(*)$ say that $X \in \overrightarrow{AB}$, and the last two that $X \in \overrightarrow{AC}$. So $\overrightarrow{AB} \cup \overrightarrow{AC} = m$, and $\overrightarrow{AB}$ and $\overrightarrow{AC}$ are by definition opposite rays.

Now B-A-C and the Unique Middle Theorem imply that $B \notin \overrightarrow{AC}$ and $C \notin \overrightarrow{AB}$. So $\overrightarrow{AB} \cap \overrightarrow{AC}$ contains neither B nor C. We consider points $X \neq A$, B, or C on m:

Suppose that $X \in \overrightarrow{AB} \cap \overrightarrow{AC}$. Then $X \in \overrightarrow{AB}$ implies that one of X-B-A or B-X-A holds, and $X \in \overrightarrow{AC}$ means that one of A-X-C or A-C-X is true. Thus, two of the four relations X-B-A, B-X-A, A-X-C, A-C-X hold. Now Theorem 9.2 implies that $\omega < \infty$ and $X = A_m^*$.

It follows that $\overrightarrow{AB} \cap \overrightarrow{AC} \subseteq \{A, A_m^*\}$. By Proposition 9.3, $\{A, A_m^*\} \subseteq \overrightarrow{AB} \cap \overrightarrow{AC}$ when $\omega < \infty$.

Corollary 9.7. Each ray has a unique opposite ray.

Proof. Let $\overrightarrow{AB}$ be a ray (so $AB < \omega$). By Proposition 8.11, there exists point C on $\overleftrightarrow{AB}$ with C-A-B. Note that $CA + AB = CB \leq \omega$ implies that $AC < \omega$. By the Opposite Ray Theorem, $\overrightarrow{AC}$ is an opposite ray to $\overrightarrow{AB}$.

Now let h be any ray opposite $\overrightarrow{AB}$, and let C be exactly as in the preceding paragraph. By definition of "opposite," A is an endpoint of h and $h \cup \overrightarrow{AB} = \overleftrightarrow{AB}$. C is on $\overleftrightarrow{AB}$, but C-A-B means that C is *not* on $\overrightarrow{AB}$. Hence $C \in h$. Since $AC < \omega$, $h = \overrightarrow{AC}$ by Theorem 8.4.

We denote the ray opposite to ray h by h'. So $\overrightarrow{AB}'$ means the ray opposite $\overrightarrow{AB}$. Corollary 9.7 says that $\overrightarrow{AB}'$ is uniquely determined by $\overrightarrow{AB}$.

Corollary 9.8. Let A and B be points on a line m with $0 < AB < \omega < \infty$. Then $\overrightarrow{AB}' = \overrightarrow{AB^*_m}$.

Proof. Problem 4.

Corollary 9.9. Let A and B be points on a line m with $0 < AB < \omega < \infty$. Then $m = \overline{AB} \cup \overline{BA^*_m} \cup \overline{A^*_m B^*_m} \cup \overline{B^*_m A}$, the interiors of these segments being disjoint.

Proof. Problem 5.

The final result in this chapter provides a nice equivalent condition for when three points on a line do *not* satisfy a betweenness relation.

Theorem 9.10. Let A and B be points on a line m with $0 < AB < \omega < \infty$. Let $C \neq A$, B, A^*_m, or B^*_m be another point on m. Then there is no betweenness relation among A, B, and C if and only if $C \in \overline{A^*_m B^*_m}^o$ (i.e., A^*_m-C-B^*_m).

Proof. Problem 6.

Problem Set 9

1. Let $\overrightarrow{AB}$ be a ray with carrier m, and C a point in $\overrightarrow{AB}^o$. Prove that if $\omega < \infty$, then $C^*_m \notin \overrightarrow{AB}$.

2. Prove that *whenever* $\omega < \infty$, then $\overrightarrow{AB} \cup \overrightarrow{BA} \neq \overleftrightarrow{AB}$.

3. Prove Proposition 9.3. (Hint: Use Theorem 9.1, the definition of ray, and the Rule of Insertion to show part (a). Then use (a) to obtain (b).)

4. Prove Corollary 9.8.

5. Prove Corollary 9.9. (Hint: Proposition 9.3 and Corollary 9.8.)

6. Prove Theorem 9.10. (Hint: First show that there is a betweenness relation among A, B, and C if and only if C is in $\overrightarrow{AB} \cup \overrightarrow{BA}$. Then use Proposition 9.3 and Corollary 9.9.)

7. Prove that if A is a point on a line l, and if s is any real number with $0 < s < \omega$, then there are exactly two points P on l such that $AP = s$.

8. Let A, B, C be distinct points on a line m and assume that $\omega < \infty$. Prove that $A_m^*\text{-}C\text{-}B_m^* \Leftrightarrow A\text{-}C_m^*\text{-}B$.

9. Prove that for any four distinct points on a line, there must exist a betweenness relation among some three of them.

10. Prove that if $\omega < \infty$ and $\overrightarrow{AB}, \overrightarrow{CD}$ are rays with $\overrightarrow{AB} \subseteq \overrightarrow{CD}$, then $\overrightarrow{AB} = \overrightarrow{CD}$.

11. Let A, B, C, D be four distinct points on a line m such that $A\text{-}B\text{-}C$ and $D \in \overrightarrow{A_m^* B_m^*}{}^o$. Find (with proof) a betweenness relation among B, C, D.

12. Assume that $\omega = 9$ and that A, B, C, X, Y are distinct points on a line m, $A\text{-}B\text{-}C$, $AB = 3$, $BC = 1$, $XB = 8$, $Y \in \overrightarrow{BA}$, and $X \notin \overrightarrow{BC}$. For each statement, answer true or false and explain why.

   (a) There is a betweenness relation among Y, A, B.

   (b) There is a betweenness relation among X, A, B.

   (c) There is a betweenness relation among B, C, X.

   (d) There is a betweenness relation among A, C, X.

   (e) $CA_m^* = 6$.

   (f) $AX + XY \geq AY$.

   (g) $\overrightarrow{AB_m^*} = \overrightarrow{AB}$.

   (h) $\overrightarrow{AC} = \overrightarrow{AB}$.

   (i) $m = \overrightarrow{AB} \cup \overrightarrow{CB}$.

   (j) There is a point P on $\overrightarrow{AB}$ with $AP = 8.57$.

   (k) There is a point Q on $\overrightarrow{AB}$ with $AQ = 9.57$.

(l) There is a ray with endpoint A that is not contained in m.

(m) $\overrightarrow{BA} \cap \overrightarrow{BC} = \{B\}$.

(n) $AB, \overline{AB}, \overrightarrow{AB}, \overleftrightarrow{AB}$ all mean the same thing.

(o) $m = \overrightarrow{AA^*_m}$.

10   Separation

Chapters 6–9 were limited to a discussion of points on a given line. Now we begin to explore properties of points and lines throughout $\mathbb{P}$.

Diagrams become important at this stage in explaining theorems and motivating proofs. But a diagram itself is never a proof. Most of the pictures that we draw show configurations in $\mathbb{E}$ at best and may not be entirely accurate or complete even there. They certainly cannot reveal all the subtlety of the general setup. Pictures are useful in motivating a path to follow in constructing a proof; that is, in bringing to mind those previously established results that might be relevant. The reader, in attempting to do a problem, should draw pictures for just this purpose. But the actual proof of a theorem is a sequence of logically justified statements, not a picture.

An important property in Euclidean plane geometry, but one that was only tacitly assumed in that setting and never explicitly assumed or proved, is the following: Each line in $\mathbb{E}$ divides the plane into two convex sets called halfplanes, and any segment that has one endpoint in each halfplane intersects the line.

To ensure that a suitable generalization of this property holds in our abstract geometry, another axiom is needed. It will be introduced later in this chapter. We begin with the definition of convexity.

Definition. A subset S of $\mathbb{P}$ is *convex* if for each pair of points $X \neq Y$ of S with $XY < \omega$, the entire segment $\overline{XY}$ lies in S. (In other words, for each $X \neq Y$ in S with $XY < \omega$, whenever X-Z-Y, then $Z \in S$. See Figure 10.1.)

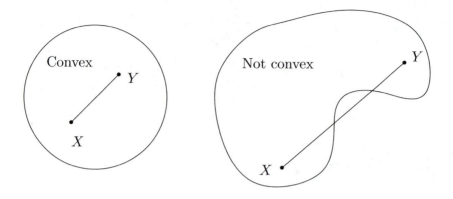

Figure 10.1

Theorem 10.1. If S_1 and S_2 are convex, then so is $S_1 \cap S_2$.

Proof. We need to show that if $X \neq Y$ are in $S_1 \cap S_2$ with $XY < \omega$, then $\overline{XY} \subseteq S_1 \cap S_2$:

X, Y in $S_1 \cap S_2$ implies that X, Y are in S_1 (as $S_1 \cap S_2 \subseteq S_1$), which implies that $\overline{XY} \subseteq S_1$, as S_1 is convex. Similarly, X, Y in $S_1 \cap S_2$ implies that X, Y are in S_2. Hence $\overline{XY} \subseteq S_2$, as S_2 is convex. Thus $\overline{XY} \subseteq S_1 \cap S_2$.

Theorem 10.2. Segments, rays, and lines are convex sets.

Proof. First, we show that any line l is a convex set: If X, Y are points on l with $XY < \omega$, then $l = \overleftrightarrow{XY}$ by Axiom I4. Then $\overline{XY} \subseteq \overleftrightarrow{XY} = l$ by Proposition 6.3.

To show that any ray $\overrightarrow{AB}$ is a convex set, we must show that for any points X, Y in $\overrightarrow{AB}$ with $XY < \omega$, then $\overline{XY} \subseteq \overrightarrow{AB}$.

If $X = A$, then $AY = XY < \omega$ implies that $\overrightarrow{AB} = \overrightarrow{AY} = \overrightarrow{XY}$ by Theorem 8.4. Now by definition of $\overrightarrow{XY}$, $\overline{XY} \subseteq \overrightarrow{XY} = \overrightarrow{AB}$. The same holds if $Y = A$.

If neither X nor Y is A, then either A-X-Y or A-Y-X, according to Theorem 8.3; and since $\overline{XY} = \overline{YX}$, we may as well suppose the former. Then $AX < AY \leq \omega$, so that $\overrightarrow{AB} = \overrightarrow{AX}$ (Theorem 8.4). If X-Z-Y, then A-X-Z-Y by insertion, and hence A-X-Z. So $Z \in \overrightarrow{AX} = \overrightarrow{AB}$. Consequently, $\overline{XY} \subseteq \overrightarrow{AB}$, and it follows that each ray is a convex set.

We have proved that any ray $\overrightarrow{AB}$, and hence $\overrightarrow{BA}$ as well, is convex. Problem 9 in Chapter 6 shows that $\overline{AB} = \overrightarrow{AB} \cap \overrightarrow{BA}$. Thus $\overline{AB}$ is convex by Theorem 10.1.

Definition. A pair of sets H, K is called *opposed around a line* m if H and K are each nonempty and convex, if they are disjoint $(H \cap K = \emptyset)$, and if $H \cup K = \mathbb{P} - m$.

For any line m in each of the five examples of Chapter 1, there exists a pair of sets opposed around m, as indicated in Figures 10.2 through 10.5.

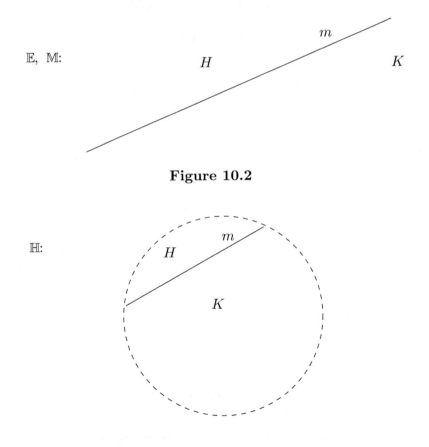

E, M:

Figure 10.2

ℍ:

Figure 10.3

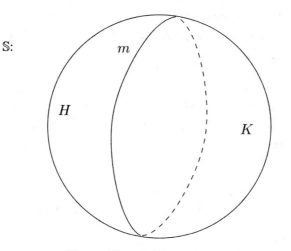

S:

Here, H and K are hemispheres.

Figure 10.4

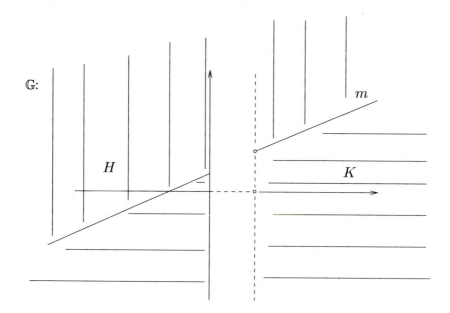

G:

Figure 10.5

Note that from the definition, a line m is disjoint from each of a pair of sets that is opposed around m. Note also that the pairs of sets H, K depicted in Figures 10.2 through 10.4 for $\mathbb{E}, \mathbb{M}, \mathbb{H}$, and $\mathbb{S}$ all share the following property:

If $X \in H$, $Y \in K$ and $XY < \omega$, then $\overline{XY} \cap m$ is not empty.

Roughly speaking, this means that we cannot cross from H to K in the given plane without meeting some point of m. This condition does *not* hold with respect to the pair H, K sketched for $\mathbb{G}$ in Figure 10.5 and in Figure 10.6.

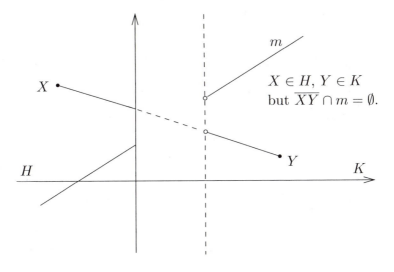

Figure 10.6

This extra condition, true for the opposed pairs of sets shown for $\mathbb{E}, \mathbb{M}, \mathbb{H}$, and $\mathbb{S}$, but not for $\mathbb{G}$, is so important that we define special terminology for those pairs of sets that satisfy it.

Definition. Let m be a line. Sets H and K are called *opposite halfplanes with edge m* if H and K are opposed around m and if, whenever X is any point of H and Y is any point of K with $XY < \omega$, then $\overline{XY} \cap m$ is not empty.

So the sets H, K sketched in Figures 10.2 through 10.4 for $\mathbb{E}, \mathbb{M}, \mathbb{H}$, and $\mathbb{S}$ are opposite halfplanes with edge the given line m. The sets H, K shown for $\mathbb{G}$ in Figure 10.6 are opposed around m but are *not* opposite halfplanes with edge m.

The existence, for any line m, of a pair of opposite halfplanes with edge m is an essential property in order that an abstract plane be two dimensional. It is not enough that each line merely has a pair of sets opposed around it. The next example illustrates this.

Example. Let $\mathbb{P} = \mathbb{R}^3$, the set of all points in three-dimensional Euclidean space. Then $\mathbb{R}^3$, with the usual Euclidean lines and distance, satisfies all of the first 11 axioms and $\omega = \infty$. Each point corresponds to a triple of coordinates (x, y, z) with respect to the usual mutually perpendicular x-, y-, and z-axes. Let m be the z-axis. That is, $m = \{(0, 0, z) : z \in \mathbb{R}\}$. Define

$$H = \{(x, y, z) : x > 0,\ y \geq 0\} \cup \{(x, y, z) : x \leq 0, y > 0\}\,;$$

$$K = \{(x, y, z) : x \geq 0, y < 0\} \cup \{(x, y, z) : x < 0, y \leq 0\}\,.$$

Then each of H and K is a half-space, bounded on one side by the xz-plane ($y = 0$). The part of this plane with $x > 0$ is contained in H and the part with $x < 0$ is in K. H and K project vertically onto the subsets of the xy-plane ($z = 0$) shown in Figures 10.7 and 10.8.

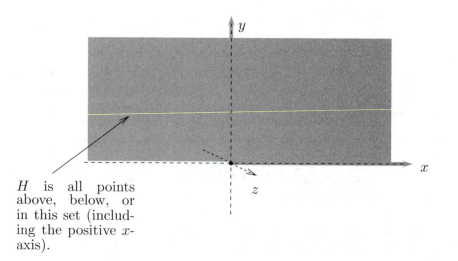

H is all points above, below, or in this set (including the positive x-axis).

Figure 10.7

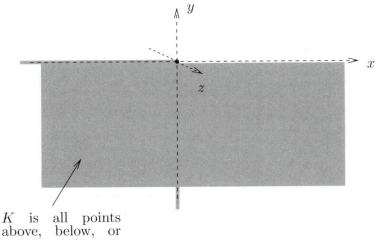

K is all points above, below, or in this set (including the negative x-axis).

Figure 10.8

It is not hard to check that H and K are opposed around m but are not a pair of opposite halfplanes with edge m. (There exist many points $X \in H$ and $Y \in K$ with $\overline{XY} \cap m = \emptyset$).

There are other pairs of sets opposed around the z-axis. One such pair is I, J, where

$$I = \left\{(x,y,z)\Big| y > 0, x \leq 0\right\} \cup \left\{(x,y,z)\Big| y \leq 0, x < 0\right\};$$

$$J = \left\{(x,y,z)\Big| y \geq 0, x > 0\right\} \cup \left\{(x,y,z)\Big| y < 0, x \geq 0\right\}.$$

Note that I and J are obtained by rotating H and K, respectively, $90°$ about the z-axis. A rotation of H and K through any fixed angle about the z-axis will produce another pair of sets opposed around m. But none of these is a pair of opposite halfplanes with edge m.

The next result describes a relationship between any pair of sets opposed around a line m and any pair of opposite rays that have a common endpoint on m but whose union is a different line.

Theorem 10.3. Let m be a line in $\mathbb{P}$ and suppose that H and K are a pair of sets opposed around m. Suppose also that A and C are points such that $C \in m$, $A \in H$ and $AC < \omega$. Then $\operatorname{Int} \overrightarrow{CA} \subseteq H$ and $\operatorname{Int} \overrightarrow{CA}' \subseteq K$.

Proof. Let $l = \overleftrightarrow{AC}$. Then $l \neq m$, since $A \notin m$. So l meets m only at C (except that if $C_l^* = C_m^*$ when $\omega < \infty$, then the lines meet at this point also). Thus $\operatorname{Int} \overrightarrow{CA}$ and $\operatorname{Int} \overrightarrow{CA}'$ are contained in $\mathbb{P} - m = H \cup K$.

Suppose that X, Y are points on $\overleftrightarrow{AC}$ such that X-C-Y and $XY < \omega$. Then one of X and Y is in H and the other is in K. For if, say, $X \in H$ and $Y \in H$, then $\overline{XY} \subseteq H$ since H is convex. But this contradicts $C \in \overline{XY}$ and $C \in m$. We shall use this fact several times in the rest of the proof.

There is a point B on $\overleftrightarrow{AC}$ with A-C-B and $AB < \omega$ (Proposition 8.11). Then $A \in H$ implies that $B \in K$. Also, $\overrightarrow{CB} = \overrightarrow{CA}'$ (9.6).

Let $X \in \operatorname{Int} \overrightarrow{CA}$. Then $0 < CX < \omega$. We wish to prove that $X \in H$. Again by Proposition 8.11, there is a point U with X-C-U and $XU < \omega$. Hence, $U \in \overrightarrow{CX}' = \overrightarrow{CA}' = \overrightarrow{CB}$.

If C-B-U, then X-C-U and Insertion yield X-C-B and $XB < \omega$. Then $B \in K$ implies that $X \in H$. If $U = B$, we again have X-C-B and $XB < \omega$, hence $X \in H$.

If C-U-B, then A-C-B and Insertion force A-C-U, with $AU < \omega$. So $A \in H$ implies that $U \in K$. Thus X-C-U and $XU < \omega$ yields once more that $X \in H$. We have proved that $\operatorname{Int} \overrightarrow{CA} \subseteq H$. The same argument for the point $B \in K$ establishes that $\operatorname{Int} \overrightarrow{CA}' = \operatorname{Int} \overrightarrow{CB} \subseteq K$.

Corollary 10.4. Let m be a line and suppose that H and K are a pair of sets opposed around m. Let A and B be two points not on m such that A-X-B for some point $X \in m$. Then A and B lie one in each of H and K, in some order.

Proof. Since $\mathbb{P} - m = H \cup K$, it suffices to assume that $A \in H$ and prove that $B \in K$. Now A-X-B implies that $XA < \omega$, $XB < \omega$, and $\overrightarrow{XB} = \overrightarrow{XA}'$ (9.6). By Theorem 10.3, $\operatorname{Int} \overrightarrow{XA} \subseteq H$ and $\operatorname{Int} \overrightarrow{XB} \subseteq K$. Therefore, $B \in K$.

The following two results require the stronger hypothesis that the given line has a pair of opposite halfplanes for which it is the edge.

Theorem 10.5. Suppose that m is a line such that there exist a pair H, K of opposite halfplanes with edge m. Suppose also that $\omega < \infty$ and $A \in m$. If B is any point with $AB = \omega$, then $B \in m$. (That is, $B = A_m^*$, so that there is only one point $B \in \mathbb{P}$ with $AB = \omega$).

Proof.   Suppose toward a contradiction that $B \notin m$. By Axioms I3 and N, there is a line n through A and B and a point P on n with $0 < AP < \omega$. Then $0 < BP < \omega$ (Theorem 9.1). Since $B \notin m$, $n \cap m = \{A\}$. Then we may choose notation for the given halfplanes to assume that $P \in H$. By Theorem 10.3, Int $\overrightarrow{AP} \subseteq H$ and Int $\overrightarrow{AP}' \subseteq K$.

There is a point Q on n with $P\text{-}B\text{-}Q$ and $PQ < \omega$ (Proposition 8.11). Thus $Q \notin \overrightarrow{BP}$; and $B = A_n^*$ implies that $\overrightarrow{BP} = \overrightarrow{AP}$ (Proposition 9.3). So $Q \in \overrightarrow{AP}'$ with $Q \neq A$ or B. Hence, $Q \in K$. Now, by definition of halfplanes H and K, $\overline{PQ}$ meets m. Since $\overline{PQ} \subseteq n$, this yields $A \in \overline{PQ}$. Now A and B are both in $\overline{PQ}$, with $AB = \omega > PQ$. This contradicts Proposition 8.7. Thus $B \in m$ and the theorem is proved.

Theorem 10.6. Suppose that m is a line such that there exists a pair H, K of opposite halfplanes with edge m. Let A and B be two distinct points not on m. Then A and B lie one in each of H and K (in some order) if and only if there is a point X on m such that $A\text{-}X\text{-}B$.

Proof.   One of the claimed implications has already been proved in Corollary 10.4. To establish the other one, we may assume that $A \in H$ and $B \in K$. If $AB < \omega$, then $\overline{AB}$ meets m, simply by definition of opposite halfplanes with edge m; hence, there is a point $X \in m$ with $A\text{-}X\text{-}B$.

Suppose that $AB = \omega$. Let l be a line through A and B (Axiom I3). By Axiom N and Theorem 9.1, there is a point P on l with $A\text{-}P\text{-}B$, hence $AP < \omega$ and $BP < \omega$. If $P \in m$, we are done. If $P \in H$, then $\overline{PB}$ meets m; if $P \in K$, then $\overline{AP}$ meets m, by the definition of opposite halfplanes. In either case, l meets m at some point X. By Theorem 9.1, $A\text{-}X\text{-}B$.

Corollary 10.7. Suppose that m is a line such that there exists a pair H, K of opposite halfplanes with edge m. Then H, K is the only pair of sets opposed around m. That is, if I, J is any pair of sets opposed around m, then I and J equal H and K in some order.

Proof. Problem 7.

The preceding examples, for three-dimensional Euclidean space $\mathbb{R}^3$ and for $\mathbb{G}$, displayed pairs of sets opposed around a line m that were not opposite halfplanes with edge m. It follows from Corollary 10.7 that in these examples the given line m cannot be the edge of a pair of opposite halfplanes. So it is impossible to prove from the axioms introduced so far that for every line m there exists a pair of opposite halfplanes with edge m. Since this property is essential for the further development of our system of plane geometry, we adopt it now as our next axiom.

Separation Axiom (Axiom S):

For each line m, there exists a pair of opposite halfplanes with edge m.

The halfplanes with edge m, usually denoted H and K, are uniquely determined by m, by Corollary 10.7. Of course, if we change to a different line, the corresponding halfplanes will change as well. Now that we are assuming the Separation Axiom, Theorems 10.3, 10.5, and 10.6 and Corollaries 10.4 and 10.7 apply to any line in $\mathbb{P}$.

We have seen that Axiom S is true for $\mathbb{E}, \mathbb{M}, \mathbb{H}$, and $\mathbb{S}$ and that it is false for $\mathbb{G}$ and for three-dimensional Euclidean space.

We next use the concept of opposite halfplanes to define the notion of "same side" or "opposite sides" of a line. It is important to remember that the only meaning these words have in our theory is in terms of halfplanes.

Definition. Let H and K be halfplanes with edge m. We say that two points that both lie in H (or both in K) lie on the *same side* of m; if one point is in H and the other is is K, we say that they are on *opposite sides* of m.

Now we derive some easy consequences of Axiom S and the previous theorems.

Theorem 10.8. Suppose that $\omega < \infty$. For each point A there is exactly one point A^* in $\mathbb{P}$ with $AA^* = \omega$. Furthermore, every line through A also passes through A^*.

Proof. Let A be any point in $\mathbb{P}$ and let m be any line through A. There is a unique point A_m^* on m with $AA_m^* = \omega$ (Theorem 9.1). Let B be any point in $\mathbb{P}$ with $AB = \omega$. Axiom S and Theorem 10.5 imply that $B = A_m^*$. Hence, B is unique and m passes through B. We denote $B = A^*$ and observe that the theorem is proved.

Definition. A^* is called the *antipode* of A.

Corollary 10.9. Suppose that $\omega < \infty$. Then for each line m and point P there are just two possibilities: P and P^* both lie on m, or P and P^* lie on opposite sides of m.

Proof. If one of P and P^* lies on m, so does the other (Theorem 10.8). Otherwise, let H be the halfplane with edge m that contains P and let K be the opposite halfplane. Let X be any point of m. Then $PX < \omega$ (10.8), so line $\overleftrightarrow{PX}$ is defined. Then $P^* \in \overleftrightarrow{PX}$ (10.8), and P-X-P^* (9.1). Thus $P^* \in K$ by Theorem 10.6.

Theorem 10.10. (Pasch's Axiom) Let A, B, and C be any three noncollinear points (so $AB, BC, AC < \omega$ by Theorem 10.8), X a point with B-X-C, and m any line through X that does not pass through A, B, or C. Then exactly one of the following holds:

(a) m contains a point Y with A-Y-C, or

(b) m contains a point Z with A-Z-B.

Proof. Problem 8.

The next result in this chapter anticipates the discussion of parallel (i.e., disjoint) lines in Chapter 17. It says that in a plane of finite diameter there is *no* pair of parallel lines.

Theorem 10.11. Suppose that $\omega < \infty$. Any two distinct lines must have a point (in fact, a pair of antipodal points) in common.

Proof. Let $n \neq m$ be lines. Suppose (toward a contradiction) that $n \cap m = \emptyset$. Let H and K be the opposite halfplanes with edge m (Axiom S). Let P be any point on n. We may assume that $P \in H$. Then $P^* \in n$ and $P^* \in K$ (Corollary 10.9). Let Q be any other point on n. Then Q is in H or K, and so either $\overline{QP^*}$ or $\overline{PQ}$ meets m, by the definition of opposite halfplanes. In either case n meets m, which is a contradiction.

Corollary 10.12. Suppose that $w < \infty$. If h is a ray and l a line in a plane $\mathbb{P}$, then exactly one of the following must hold:

(i) $h \subseteq l$;

(ii) $h \cap l = \{A, A^*\}$, where these are the endpoints of h;

(iii) $h \cap l$ is a single interior point of h.

Proof. Problem 9.

Problem Set 10

1. Assume $w < \infty$. Show that if A^* is the antipode of A and B is any other point, then $A\text{-}B\text{-}A^*$ and $BA^* = w - AB$.

2. Assume $w < \infty$. Let A and B be any two points with $AB < w$. Show that $A^*B^* = AB$. (Hint: Problem 1 is relevant.)

3. Prove: If $w < \infty$, then $A^{**} = A$ for all points A, and $A\text{-}B\text{-}C$ if and only if $A^*\text{-}B^*\text{-}C^*$.

4. Show that in the example $\mathbb{P} = \mathbb{R}^3$ presented on page 106 of this chapter, the sets H and K are indeed a pair of sets opposed around the z-axis but are not a pair of opposite halfplanes with edge the z-axis.

5. Suppose P is a point not on a line $m = \overleftrightarrow{AB}$, and suppose X and Y are points with $A\text{-}X\text{-}P$ and $P\text{-}B\text{-}Y$. Show that $XY < w$ and that $\overline{XY}$ meets m.

6. Let m be a line and P, Q points such that $P \notin m$, $PQ = 1$, and $PX \geq 2$ for all X on m. Prove that P and Q lie on the same side of m.

7. Prove Corollary 10.7. (Hint: Since $H \cup K = \mathbb{P} - m = I \cup J$, we may assume that there is some point $A \in H \cap I$ and then prove that $K = J$ and $H = I$. First use Theorem 10.6 and Corollary 10.4 to prove $K \subseteq J$. Note that Axiom S is not to be used.)

8. Prove Theorem 10.10.

9. Prove Corollary 10.12.

10. Let m, n be distinct lines. Let H, K be the opposite halfplanes with edge m and I, J the opposite halfplanes with edge n. Prove that $m \cap n \neq \emptyset$ if and only if $H \cap I \neq \emptyset$ and $H \cap J \neq \emptyset$.

11   Pencils and Angles

In this chapter we introduce a new undefined term and seven new axioms, to provide for the arrangement of rays about a point. Our intention is to set things up in the abstract theory so shrewdly that the sequencing of rays about a point in any plane corresponds precisely to the sequencing of points on a line in a plane of finite diameter. This correspondence, which we call a duality, will be so perfect that we will not have to prove many of our results, because their proofs correspond precisely to proofs we have already given.

Once this notion of duality has been clarified, we shall be well equipped to define angles and establish their basic properties.

Pencils and Angular Distance.   We begin by explaining the terms "coterminal" and "pencil," and then we introduce the angular distance between two coterminal rays.

Definition. Two rays with the same endpoint are called *coterminal*. The *pencil of rays at* A is the set of all rays with endpoint A. We denote this set by $\boldsymbol{P}$ or $\boldsymbol{P}_A$. (See Figure 11.1.)

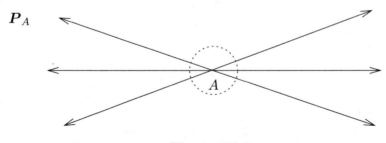

$\boldsymbol{P}_A$

Figure 11.1

Note that when $\omega < \infty$, each ray $h = \overrightarrow{AB}$ has two endpoints, A and A^*; and $h = \overrightarrow{AB} = \overrightarrow{A^*B}$. Thus, as in Figure 11.2,

$$\boldsymbol{P}_A = \boldsymbol{P}_{A^*} \qquad \text{when } \omega < \infty.$$

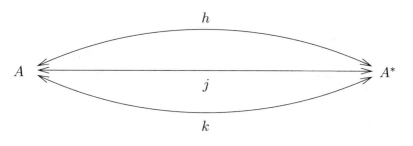

Figure 11.2

To provide for the arrangement of rays in a pencil requires seven axioms and a new undefined term, a real-valued function μ that assigns to each two coterminal rays p, q a real number $\mu(p,q)$, called the *angular distance* between p and q. We agreed earlier to write PQ for the distance $d(P,Q)$ between two points P and Q, to simplify the notation; and we agree now to do the same thing with the angular distance between two rays: We shorten $\mu(p,q)$ to pq. As a standing notation we continue to write h' for the ray opposite a given ray h.

We introduce four new axioms to set the basic properties of the angular distance function.

Axioms of Angular Distance: For all coterminal rays p and q,

M1. (Positivity) $0 \le pq \le 180$.

M2. (Definiteness) $pq = 0$ if and only if $p = q$.

M3. (Symmetry) $pq = qp$.

M4. (Opposites) $pq = 180$ if and only if $q = p'$.

If we think of pq as the familiar measure of the angle formed by the coterminal rays p and q in the coordinate plane $\mathbb{E}$, the reasonableness of these assumptions is entirely clear. Axiom M3 requires that the angular distance between two coterminal rays be independent of the order in which the two rays are listed. Axiom M1 says that the range of the function $\mu(p, q)$ lies in the closed real interval $[0, 180]$. Axiom M2 says that the lower equality holds precisely when $p = q$, and Axiom M4 requires that the upper equality hold precisely when p and q are opposite rays.

To define betweenness for coterminal rays, we parallel the path we followed in Chapter 6 when we investigated betweenness for points on a line.

Definition. Ray b lies *between* rays a and c (written a-b-c) provided that

(a) a, b, c are different, coterminal rays, and

(b) $ab + bc = ac$.

Theorem 11.1 a-b-$c \Leftrightarrow c$-b-a.

Proof. Assume first that a-b-c. The rays c, b, and a are different, coterminal rays, because rays a, b, and c are different and coterminal. Furthermore,

$$
\begin{aligned}
cb + ba &= bc + ab \text{ (by Axiom M3)} \\
&= ab + bc \text{ (by commutativity of addition in } \mathbb{R}) \\
&= ac \text{ (because } a\text{-}b\text{-}c) \\
&= ca \text{ (by Axiom M3)}.
\end{aligned}
$$

Therefore, the definition is satisfied, and c-b-a. We have proved that a-b-$c \Rightarrow c$-b-a. Since this result holds for *any* three rays that satisfy a betweenness relation, it follows that c-b-$a \Rightarrow a$-b-c as well.

If you think that this argument seems familiar, you are correct. It is precisely the same argument we used earlier to prove the corresponding result for points on a line. This is our first instance of the duality mentioned previously, a notion whose systematic exploitation will save us considerable labor.

The sequencing of points on a line in a plane of finite diameter is governed by three additional axioms, Betweenness of Points (BP), Quadrichotomy for Points (QP), and the Real Ray Axiom (RR).

We select analogous axioms for rays in a pencil with the view of preserving the analogy.

Betweenness of Rays Axiom (Axiom BR):

If a, b, and c are different, coterminal rays and if $ab + bc \leq 180$, then there exists a betweenness relation among a, b, and c.

Quadrichotomy of Rays Axiom (Axiom QR):

If a, b, c, x are distinct coterminal rays and a-b-c, then at least one of the following must hold: x-a-b, a-x-b, b-x-c, or b-c-x.

We use the terms "wedge" and "fan" for the subsets of a pencil that are analogous to "segment" and "ray." To be precise, we make the following definition.

Definition. Let p, q be coterminal rays with $0 < pq < 180$. The *wedge* $\overline{pq}$ (pictured in Figure 11.3) and *fan* $\overrightarrow{pq}$ (pictured in Figure 11.4) are defined to be

$$\overline{pq} = \{p, q\} \cup \{r : p\text{-}r\text{-}q\},$$
$$\overrightarrow{pq} = \{p, q\} \cup \{r : p\text{-}r\text{-}q\} \cup \{r : p\text{-}q\text{-}r\} = \overline{pq} \cup \{r : p\text{-}q\text{-}r\}.$$

We agree not to use the notations $\overline{pq}$ and $\overrightarrow{pq}$ unless $0 < pq < 180$.

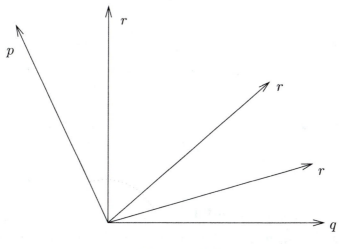

Figure 11.3

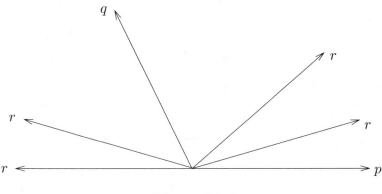

Figure 11.4

Figure 11.4 suggests that the set of all points on all rays in a fan (except for those on the "border") ought to coincide with a halfplane as described in Chapter 10. This is true, in fact, and is proved in Theorem 12.2 of Chapter 12.

The analog for coterminal rays of the nontriviality axiom (Axiom N) for collinear points need not be included as an axiom because it can be proved as a theorem.

Theorem 11.2 (Nontriviality) For any ray p there is a coterminal ray q with $0 < pq < 180$.

Proof. Let m be the carrier of p, and (by Theorem 7.6) take any point $C \notin m$. If A is an endpoint of p, then $AC < \omega$ (otherwise $C \in m$ by Theorem 10.5). Then the ray $q = \overrightarrow{AC}$ is coterminal with p, and $0 < pq < 180$ by Axioms M1, M2, and M4.

In particular, fans exist in our abstract plane. Our final axiom for rays in a pencil is the analog for coterminal rays of the Real Ray Axiom (Axiom RR) (see Figure 11.5).

Real Fan Axiom (Axiom RF):

For any fan $\overrightarrow{pq}$ and any real number t with $0 \le t \le 180$, there is a ray r in $\overrightarrow{pq}$ with $pr = t$.

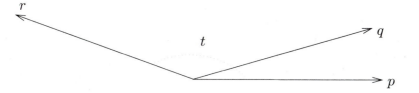

Figure 11.5

There is nothing magical about the number 180 in all these axioms. We choose it because it gives us the familiar degree measure for angles. Any other positive real number would do as well; we would simply be using different units to measure angular distances. (The choice of 1/2, for example, would mean that we have chosen to use revolutions, the choice of 2 would mean we are using right angles or quarter-turns, and the choice of π would mean radians. We use degrees mainly in keeping with a tradition that dates back at least 3500 years to the ancient Egyptians and Babylonians.)

The Angular Distance Function in the Five Models.   We need to introduce a specific angular distance function μ for each of our five major examples.

$\mathbb{E}$:   For $\mu(p, q)$ in the coordinate plane $\mathbb{E}$ we take the familiar measure in degrees of the angle between the coterminal rays p and q (Figure 11.6).

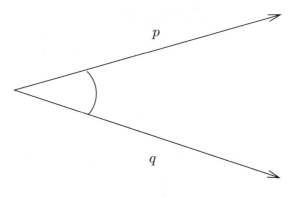

Figure 11.6

$\mathbb{M}$:   Rays in the Minkowski plane are exactly the same as those in the Euclidean plane $\mathbb{E}$, so we use the same angular distance function μ here that we use in $\mathbb{E}$.

G: Rays in the gap plane are the same as rays in the Euclidean plane E, except that some have gaps and look strange (see Figure 11.7). So again we use precisely the same notion of angular distance in G as we use in E and in M.

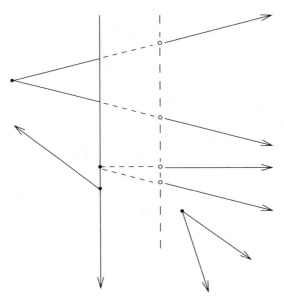

Figure 11.7

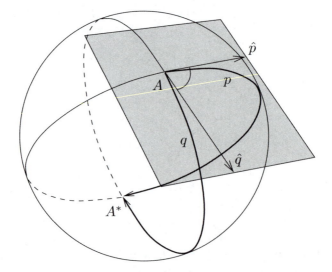

Figure 11.8

$\mathbb{S}$:   Two coterminal rays p and q in the spherical plane $\mathbb{S}$ are great semicircles that join two antipodal points, say A and A^*. The tangent plane to the sphere at A is perpendicular to the diameter of the sphere through A, and there are unique rays $\hat{p}, \hat{q}$ in that tangent plane that are tangent to the semicircles p and q, respectively, at A. We take $\mu(p, q)$ to be the angular distance between these two rays in the tangent plane (see Figure 11.8). Note that arguing at the antipodal point A^* instead of A leads to precisely the same result for $\mu(p, q)$.

$\mathbb{H}$:   Defining $\mu(p, q)$ in the hyperbolic model $\mathbb{H}$ is a little more difficult, and we settle for a brief description of the procedure. Regard the hyperbolic plane as the equatorial plane of a unit sphere (Figure 11.9). Then each chord of the equator is the vertical projection of a semicircle of the upper hemisphere into the equatorial plane, and each ray in $\mathbb{H}$ on that chord is the vertical projection of a circular arc of that semicircle. If rays p and q in $\mathbb{H}$ are coterminal at A, we take $\mu(p, q)$ to be the ordinary angle at the point on the hemisphere above A between the two tangent lines to the circular arcs that project onto p and q.

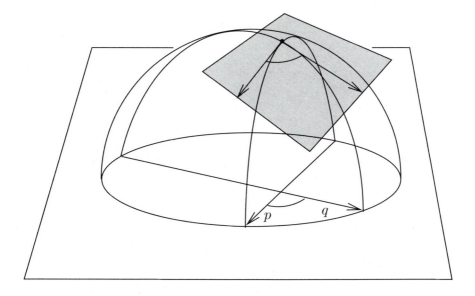

Figure 11.9

A formula for calculating $\mu(p, q)$ is derived in Proposition A.1 of Appendix II. In part, it reads as follows:

Let $p = \overrightarrow{PQ}$ and $q = \overrightarrow{PR}$, where $P = (x_0, y_0)$, $Q = (x_1, y_1)$, $R = (x_2, y_2)$ are noncollinear points in $\mathbb{H}$. Assume that $x_0 < x_1$, $x_0 < x_2$, that $\overleftrightarrow{PQ}$ has the equation $y = mx + b$ (as a Euclidean line) and that $\overleftrightarrow{PR}$ has the equation $y = nx + c$, as in Figure 11.10. Then

$$\mu(p, q) = \cos^{-1}\left(\frac{1 + mn - bc}{\sqrt{1 + m^2 - b^2}\sqrt{1 + n^2 - c^2}}\right) \text{ (in degrees)}.$$

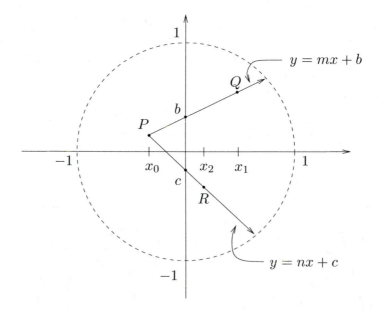

Figure 11.10

This usually differs from the ordinary Euclidean angular distance pq, which equals $\cos^{-1}((1 + mn)/\sqrt{1 + m^2}\sqrt{1 + n^2})$. (This formula can be derived from applying the Law of Cosines to the triangle PQR). See Example A.1 and Corollary A.2 of Appendix II for more specific examples.

It might seem to be more natural to take $\mu(p, q)$ to be the ordinary angular distance in $\mathbb{E}$ between the Euclidean lines that contain the rays. If we did this, the resulting plane would satisfy all of the axioms to date, but it would *not* satisfy the congruence axiom SAS, which we will add to the system in Chapter 13.

We accept as obvious the truth of the seven axioms M1–M4, BR, QR, and RF without further ado for each of the models $\mathbb{E}$, $\mathbb{M}$, $\mathbb{G}$, $\mathbb{S}$, and $\mathbb{H}$. The planes $\mathbb{E}$, $\mathbb{M}$, $\mathbb{S}$, and $\mathbb{H}$ are models of the abstract axiom system of the first 19 axioms we have introduced to date. The gap plane $\mathbb{G}$ is not a model, because, as we noted earlier, the Separation Axiom S is not satisfied.

Duality. Let us pursue the analogy we noted earlier between the betweenness theory for points on a line in a plane of finite diameter and the betweenness theory for rays in a pencil.

First, note that the basic notions and fundamental definitions parallel each other perfectly:

Line	$\longleftrightarrow$	Pencil
Collinear points	$\longleftrightarrow$	Coterminal rays
Betweenness for points A-B-C	$\longleftrightarrow$	Betweenness for rays a-b-c
Definition of segment $\overline{AB}$	$\longleftrightarrow$	Definition of wedge $\overline{ab}$
Definition of ray $\overrightarrow{AB}$	$\longleftrightarrow$	Definition of fan $\overrightarrow{ab}$
Antipodal point	$\longleftrightarrow$	Opposite ray

Any meaningful phrase about points on a line can be translated into a corresponding "dual" phrase about rays in a pencil by simply substituting the appropriate words. To obtain the dual, we just replace "point" by "ray," "distance" by "angular distance," "segment" by "wedge," "ray" by "fan," "antipodal" by "opposite," "ω" by "180," and so on. For example, the dual of "three collinear points" is "three coterminal rays"; the dual of "If $0 < AB < \omega$, then A-B-A^*" is "If $0 < ab < 180$, then a-b-a'."

The betweenness theory for points on a line is governed by just seven axioms: D1, D2, D3, BP, QP, N, and RR. The dual statements for rays in a pencil are

Axiom D2	$\longleftrightarrow$	Axiom M2
Axiom D3	$\longleftrightarrow$	Axiom M3
Axiom BP	$\longleftrightarrow$	Axiom BR
Axiom QP	$\longleftrightarrow$	Axiom QR
Axiom N	$\longleftrightarrow$	Theorem 11.2
Axiom RR	$\longleftrightarrow$	Axiom RF

Only an investigation into the dual of Axiom D1 remains. The definition of ω and Axiom D1 combine to give the true statement

$$0 \le PQ \le \omega \text{ for all collinear points } P \text{ and } Q,$$

whose dual is precisely Axiom M1, where 180 replaces and is analogous to ω. So the dual of every axiom concerning the betweenness of points on a line (in a plane of diameter $\omega < \infty$) is valid (an axiom or a theorem) in our abstract system concerning the betweenness of rays in a pencil. It follows that *the dual of any theorem about betweenness for collinear points in a plane of diameter $\omega < \infty$ is a theorem about betweenness for coterminal rays in any plane.* Indeed, the dual of the proof of the result is a proof of the dual result.

So the following theorems are all *free*, the duals of results we have proved earlier. We list them here for reference. (Some of the proofs may be assigned as problems.)

Theorem 11.3. (Dual of Theorem 6.2) If a-b-c, then b-a-c and a-c-b are both false.

Proof.   If a-b-c and b-a-c, then $ab + bc = ac$ and $ba + ac = bc$. Substituting the second into the first and using $ba = ab$ gives $ab + (ab + ac) = ac$, so that $ab = 0$, contrary to Axiom M2. The case of a-c-b is similar. (Is this argument the dual of the argument you gave for Problem 1 of Problem Set 6?)

It is easy to extend betweenness to four rays in a pencil.

Definition.   If $a, b, c,$ and d are rays, then a-b-c-d means that *all four* of a-b-c, a-b-d, a-c-d, and b-c-d hold true.

Proposition 11.4. (Dual of Proposition 7.2) If a-b-c-d, then $a, b, c,$ and d are different, coterminal rays, and d-c-b-a.

Theorem 11.5. (Dual of Theorem 7.4: Rule of Insertion for Rays)

(a) If a-b-c and a-r-b, then a-r-b-c.

(b) If a-b-c and b-r-c, then a-b-r-c.

Theorem 11.6. (Dual of Theorem 8.6: Unique Angular Distances for Fans) For any fan $\overrightarrow{pq}$ and any real number t with $0 \le t \le 180$, there is a *unique* ray r in $\overrightarrow{pq}$ so that $pr = t$. And $r \in \overline{pq}$ if and only if $t \leq pq$.

In particular, the range of the angular distance function μ is the entire closed interval $[0, 180]$.

Proposition 11.7. (Dual of Part of Proposition 8.7) If rays a and b are in a wedge $\overline{pq}$, then $ab \leq pq$.

Theorem 11.8. (Dual of Part of Theorem 9.1) If a ray a lies in a pencil $\boldsymbol{P}$, then $a\text{-}r\text{-}a'$ for every other ray $r \in \boldsymbol{P}$.

Theorem 11.9. (Dual of Theorem 9.2: Uniqueness of Quadrichotomy for Rays) If $a, b, c,$ and r are different coterminal rays and $a\text{-}b\text{-}c$, then exactly one of the following four alternatives holds: $r\text{-}a\text{-}b$, $a\text{-}r\text{-}b$, $b\text{-}r\text{-}c$, $b\text{-}c\text{-}r$, with the solitary exception that both $r\text{-}a\text{-}b$ and $b\text{-}c\text{-}r$ are correct when $r = b'$.

Theorem 11.10. (Dual of Theorem 9.6) Let p, q, r be rays in a pencil $\boldsymbol{P}$ such that $q\text{-}p\text{-}r$. Then $\overrightarrow{pq} \cup \overrightarrow{pr} = \boldsymbol{P}$ and $\overrightarrow{pq} \cap \overrightarrow{pr} = \{p, p'\}$.

Corollary 11.11. (Dual of Corollary 9.8) If p, q are rays in a pencil $\boldsymbol{P}$ with $0 < pq < 180$, then $\overrightarrow{pq} \cup \overrightarrow{pq'} = \boldsymbol{P}$ and $\overrightarrow{pq} \cap \overrightarrow{pq'} = \{p, p'\}$.

Theorem 11.12. (Dual of Theorem 9.10) Let p, q be rays in a pencil $\boldsymbol{P}$ with $0 < pq < 180$. Let $r \neq p, q, p'$, or q' be another ray in $\boldsymbol{P}$. Then there is no betweenness relation among p, q and r if and only if $p'\text{-}r\text{-}q'$.

The Compatibility Axiom. If $A\text{-}B\text{-}C$ on a line m and if $X \notin m$, must the rays $a = \overrightarrow{XA}$, $b = \overrightarrow{XB}$, and $c = \overrightarrow{XC}$ be in the same order in $\boldsymbol{P}_X$ (see Figure 11.10)? A little thought turns up nothing in the axiom system so far that might imply $a\text{-}b\text{-}c$, and yet we clearly want for this to be correct. (Indeed, it is possible to build an example of a plane that satisfies all of the axioms to date for which this property is not correct; see Problem 9 at the end of this chapter.) What is required is another axiom that we first state and then depict in Figure 11.11.

Compatibility Axiom (Axiom C):

Let $A, B,$ and C be points on a line m, let X be a point not on m, and let $a = \overrightarrow{XA}$, $b = \overrightarrow{XB}$, and $c = \overrightarrow{XC}$. If $A\text{-}B\text{-}C$, then $a\text{-}b\text{-}c$.

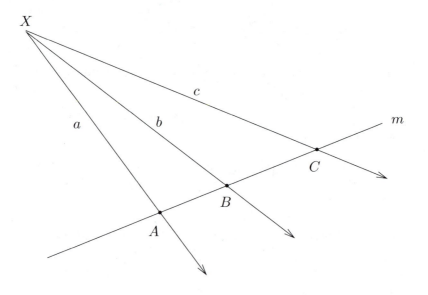

Figure 11.11

It is clear that this axiom is correct in all of the five basic models: $\mathbb{E}$, $\mathbb{M}$, $\mathbb{G}$, $\mathbb{S}$, and $\mathbb{H}$.

What about the converse implication: If a-b-c, must A-B-C follow? As very often happens in geometry, this converse is a consequence of the direct implication. We prove it next.

Theorem 11.13. (Converse of Axiom C) Let A, B, and C be points on a line m, let X be a point not on m, and let $a = \overrightarrow{XA}$, $b = \overrightarrow{XB}$, and $c = \overrightarrow{XC}$. If a-b-c, then A-B-C.

Proof. The argument is by contradiction. Suppose that a-b-c, but A-B-C is false. Then B-A-C, A-C-B, or there is no betweenness relation among A, B, C. We consider the possibilities.

Suppose that B-A-C or A-C-B. Then according to Axiom C, either b-a-c or a-c-b, and by Theorem 11.3, each contradicts the hypothesis that a-b-c.

Suppose that there is no betweenness relation among A, B, and C. Then $\omega < \infty$, and $C \neq A^*$ (otherwise A-B-C) so that $0 < AC < \omega$. It follows (Theorem 9.10) that A^*-B-C^*. Now $a' = \overrightarrow{XA^*}$ and $c' = \overrightarrow{XC^*}$ because A-X-A^* and C-X-C^*, so Axiom C implies that a'-b-c'. But then Theorem 11.12 says that no betweenness relation holds among

a, b, and c, contrary to the hypothesis that a-b-c.

Alternatively, we could have assumed the assertion of Theorem 11.13 as our axiom, in which case the present statement of Axiom C would become a theorem. See Problem 7.

Angles. The notion of "angle" is crucial for our future work, but it comes rather late in our development because the sequencing of rays in a pencil is more fundamental. We are finally ready to explain the meaning of the term. There are several different elementary notions of angle in geometry, useful for different purposes. For example, in trigonometry we use an ordered (or sensed) angle, with an initial and a terminal side and a winding number that tells how many times to go around the vertex. Useful though such an angle is in trigonometric problems, it is not needed in elementary geometry. All we need are two rays.

Definition. By an *angle* we mean the union of two coterminal rays, the *sides* of the angle. A *vertex* of an angle is a common endpoint of its sides. If rays a, b are coterminal, we adopt the notation $\underline{ab} = a \cup b$ for the angle with sides a and b. The real number ab (the angular distance, as previously defined) is the *measure* of the angle $\underline{ab}$. If $a = \overrightarrow{BA}$ and $b = \overrightarrow{BC}$, we write $\underline{\angle ABC}$ as an alternative for $\underline{ab}$ and $\angle ABC$ for ab, so that $\underline{\angle ABC} = \overrightarrow{BA} \cup \overrightarrow{BC}$. We sometimes call $\underline{ab}$ the *angle included between* the sides a and b.

We agree to restrict the notation $\angle ABC$ to the case in which $0 < BA < \omega$ and $0 < BC < \omega$, so that the sides $\overrightarrow{BA}$ and $\overrightarrow{BC}$ are defined; and we sometimes shorten $\underline{\angle ABC}$ to $\underline{\angle B}$ and $\angle ABC$ to $\angle B$ if there is no possibility of confusion. But note that an *angle* (the set) is *always* underscored, while its *measure* (the *real number*) is never underscored.

Definition. An angle $\underline{ab}$ is *zero, acute, right, obtuse,* or *straight* accordingly as $ab = 0$, $0 < ab < 90$, $ab = 90$, $90 < ab < 180$, or $ab = 180$. Angle $\underline{ab}$ is *proper* if $0 < ab < 180$.

As we have seen, a ray in a plane of infinite diameter has exactly one endpoint, so an angle in a plane of infinite diameter has a unique vertex. In a plane of finite diameter, a ray has exactly *two* endpoints, and so an angle in a plane of finite diameter has *two* vertices, which are antipodal. This is one of the assertions of the following theorem.

Proposition 11.14. (a) If $\omega < \infty$, then $\angle ABC = \angle AB^*C$.

(b) If $P \in \overrightarrow{BA}^o$ and $Q \in \overrightarrow{BC}^o$, then $\angle PBQ = \angle ABC$.

Proof. (a) According to Proposition 9.3, $\angle ABC = \overrightarrow{BA} \cup \overrightarrow{BC}$
$= \overrightarrow{B^*A} \cup \overrightarrow{B^*C} = \angle AB^*C$.

(b) According to Theorem 8.4, $\angle ABC = \overrightarrow{BA} \cup \overrightarrow{BC}$
$= \overrightarrow{BP} \cup \overrightarrow{BQ} = \angle PBQ$.

We conclude this chapter with two useful results, both of which are duals of earlier results and so do not require new proofs.

Proposition 11.15. (Dual of Proposition 8.9) If pq is an angle, there is exactly one ray b in the wedge $\overline{pq}$ so that $pb = \dfrac{1}{2}pq$.

Definition.   The ray b whose existence and uniqueness have just been noted is called the *bisector* of angle pq.

Theorem 11.16. (Dual of Theorem 9.5: Doubling an Angle) If pq is an acute angle, there is exactly one ray r in the fan $\overrightarrow{pq}$ so that q is the bisector of pr.

Problem Set 11

1. If pq is a given proper angle, show that there is a ray r such that $pr = \dfrac{1}{3}pq$ and $p\text{-}r\text{-}q$. (So each proper angle has a trisector.)

2. Prove that if rays p and q are coterminal, then $p'q' = pq$. (See Problem 2 in Chapter 10.)

3. Does the dual of the rule of two (see Problem 4 in Chapter 8) hold for coterminal rays? Either prove it or give a counterexample.

4. Show that if points A, B, C, D lie on a line m and if a point X is not collinear with A and D, then $\overrightarrow{XA}\text{-}\overrightarrow{XB}\text{-}\overrightarrow{XC}\text{-}\overrightarrow{XD} \Leftrightarrow A\text{-}B\text{-}C\text{-}D$.

5. Let A be a point not on line m. Let b, c, d be three distinct rays in $\boldsymbol{P}_A$ that meet m in distinct points B, C, D respectively, such that $C \in \overline{BD}$. Prove that $bd = bc + cd$.

6. Assume $\omega < \infty$. Suppose that A, B, C are points with $AC < \omega$ and $A\text{-}B\text{-}C$. Let X be any point not on $\overleftrightarrow{AC}$ and let A^* be the antipode of A. Prove that $\overrightarrow{XB}\text{-}\overrightarrow{XC}\text{-}\overrightarrow{XA^*}$.

7. Suppose that instead of postulating Axiom C, we assume the statement of Theorem 11.13. Prove that the statement of Axiom C will follow as a theorem. (Hint: Proceed by "dualizing" the proof of Theorem 11.13. Thus you easily can reduce to the case where there is no betweenness relation among r, s, t. Then, *provided* $\omega < \infty$ (necessary for the existence of R^* and T^*), you can dualize the rest of the proof.)

   So you may assume that $\omega = \infty$ and that $r'\text{-}t\text{-}s'$, $r'\text{-}s\text{-}t'$ and hence $r\text{-}s'\text{-}t$ (by Theorem 11.12 and Problem 2). Let X be the common endpoint of r, s, t, and choose point U with $S\text{-}X\text{-}U$ (Proposition 8.11). Then $U \in s'$ and S, T, U are noncollinear (why?). Let line $l = r \cup r'$. By Pasch's Theorem, l meets $\overline{UT}$ or $\overline{ST}$. But if l meets $\overline{ST}$, show that l meets m in two different points, which contradicts Axiom I4 and $\omega = \infty$. Finally, show that if $l(r$ or $r')$ meets $\overline{UT}$, then the hypothesis and Theorem 6.2 yield a contradiction.

8. Let p be a ray in a pencil $\boldsymbol{P}$, and let t be any real number with $0 < t < 180$. Prove that there exist exactly two rays q in $\boldsymbol{P}$ with $pq = t$. (See Problem 7 in Chapter 9.)

9. We show that Axiom C really is independent of the previous axioms by constructing a model, due to David Kay, in which the first 19 axioms hold but Axiom C fails:

   To define the new model, we modify the Euclidean model $\mathbb{E}$ by changing the way angular distance is defined on the pencil $\boldsymbol{P}_O$ of rays at the origin $O = (0, 0)$, leaving the rest of the structure undisturbed. Define a function $g_O(r)$ on $\boldsymbol{P}_O$ by taking $g_O(r)$ to be the usual (signed) trigonometric angular distance from the positive x-axis to the ray r (in the range $-180 < g_O(r) \leq 180$) *except* for the four particular rays h, k, h', k' at the (usual) angles $30, 60, -150, -120$, respectively. For these four special rays, we define $g_O(h) = 60$, $g_O(k) = 30$, $g_O(h') = -120$, and $g_O(k') = -150$. Now we use g_O to *define* the angular distance μ_O between two rays in $\mathbf{P}_O$, in the following way:

$$\mu_O(p, q) = pq = \min \{|g_O(p) - g_O(q)|,\ 360 - |g_O(p) - g_O(q)|\}\,.$$

(a) Show that all of the first 19 axioms hold, so Kay's plane is a model of the axiom system up to Axiom C.

(b) Show that if a, b, c are the rays in $\mathbf{P}_O$ through the points $A = (1, 0)$, $B = (1, \sqrt{3}/3)$, and $C = (1, \sqrt{3})$, respectively, then A-B-C, but a-b-c is false.

10. Use the formula stated just before Figure 11.10 (and proved in Appendix II) to compute $\angle QPR$ in $\mathbb{H}$, where $P = (0, \frac{1}{2})$, $Q = (\frac{1}{2}, \frac{1}{2})$, $R = (\frac{1}{2}, 0)$. How does this compare with $\angle QPR$ in $\mathbb{E}$?

12　The Crossbar Theorem

We show here that two previously introduced concepts, halfplane from Chapter 10 and fan from Chapter 11, are very closely related (as you might expect). This is done explicitly in Theorem 12.2. Next comes a corollary that is useful for finding angles with a given side and a given measure (Corollary 12.3). Then we prove the Crossbar Theorem (12.4), our main tool in determining when a ray meets a line segment. Finally, we define the notion of *interior* of an angle and present some of its fundamental properties.

Theorem 12.1. Let line l be the carrier of ray h with endpoint A. Let j and k be two rays with endpoint A, distinct from h and h', so that $h\text{-}j\text{-}k$. Then all interior points of j and k lie on the same side of l (Figure 12.1).

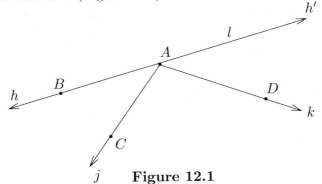

j　**Figure 12.1**

Proof. Let B, C, D be interior points of h, j, k, respectively. So $h = \overrightarrow{AB}$, $j = \overrightarrow{AC}$, $k = \overrightarrow{AD}$ (Theorem 8.4). Since j and k are not equal to either of h or h', we know that C and D are not on l. Also, $h\text{-}j\text{-}k$ implies that $0 < jk < 180$, so $\underline{jk}$ is a proper angle.

Hence $CD < \omega$. (Otherwise, C, A, and D would be collinear (Theorem 10.5), which would force j and k onto the same line, a contradiction.)

Theorem 10.3 says that all points of j^o are on the same side of l as C, and all points of k^o are on the same side of l as D. So *it suffices to prove that C and D are on the same side of l.*

Suppose not: Then C and D are on opposite sides of l. So Axiom S implies that $\overline{CD}$ meets l in some point X; and $X \neq A$ or A^*, as $\angle CAD$ is proper. Hence C-X-D, where X is an interior point of either h or h'. Thus $\overrightarrow{AX} = h$ or h' (Figure 12.2).

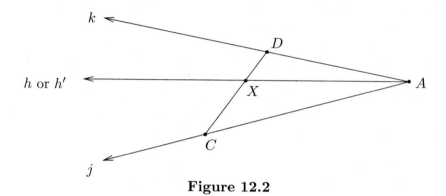

Figure 12.2

Axiom C and C-X-D imply $\overrightarrow{AC}$-$\overrightarrow{AX}$-$\overrightarrow{AD}$. Hence, either j-h-k or j-h'-k. But j-h-k contradicts h-j-k and Theorem 11.3. Furthermore, h-j-k, h-k-h' (Theorem 11.8) and Insertion for rays (Theorem 11.5) yield h-j-k-h'. Thus j-k-h', which contradicts j-h'-k and Theorem 11.3. This proves the result.

Theorem 12.2. Let l be a line and B a point in a halfplane H with edge l. Let X, A be points on l with $0 < AX < \omega$, $h = \overrightarrow{XA}$ and $k = \overrightarrow{XB}$ (Figure 12.3).

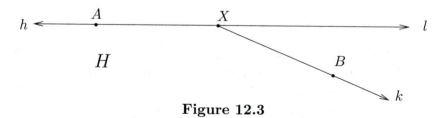

Figure 12.3

Then H consists of all points on all rays of the fan $\overrightarrow{hk}$ *except* for the points of l. In other words, $P \in H$ if and only if $P \in j^o$, for some $j \in \overrightarrow{hk}$ with $j \neq h$ or h'.

Proof. Let S be the set of all points on all rays in $\overrightarrow{hk}$, except for those on l. All points of k^o are in H, by Theorem 10.3. So if j is any ray with $j \neq h'$, and either h-j-k or h-k-j, then $j^o \subseteq H$, by Theorem 12.1. Thus $S \subseteq H$.

By Theorem 10.3, the interior points of k' are in K, the halfplane with edge l opposite H. Let T be the set of all points on all rays in $\overrightarrow{hk'}$, except for those on l. Then Theorem 12.1 again gives $T \subseteq K$.

For any point Q not on l, $XQ < \omega$ (Theorem 10.5) implies that ray $\overrightarrow{XQ}$ exists in pencil $\boldsymbol{P}_X$. By Corollary 11.11, $\overrightarrow{XQ}$ is in either $\overrightarrow{hk}$ or $\overrightarrow{hk'}$. Hence Q is in either T or S. So we have

$$T \cup S = \mathbb{P} - l = H \cup K,$$

with $S \subseteq H, T \subseteq K$, and $K \cap H = \emptyset$. It follows that $S = H$ and $T = K$. (If there were some point Y in H but not in S, then $Y \in H \cup K = S \cup T$ would force $Y \in T$. But then $T \subseteq K$ would imply $Y \in K$, which contradicts $H \cap K = \emptyset$.)

Corollary 12.3.   Let z be any number with $0 < z < 180$. For any ray $\overrightarrow{AB}$, there are exactly two rays h, k in $\boldsymbol{P}_A$ so that $\overrightarrow{ABh} = z = \overrightarrow{ABk}$. Furthermore, h^o and k^o lie in opposite halfplanes with edge $\overleftrightarrow{AB}$.

Proof. Let r be any ray in $\boldsymbol{P}_A$ with $r \neq \overrightarrow{AB}$ or $\overrightarrow{AB}'$. Thus $0 < \overrightarrow{ABr} < 180$ by Axioms M2, M4. The Unique Angular Distances for Fans Theorem says that there is a unique ray h in $\overrightarrow{ABr}$ with $\overrightarrow{ABh} = z$ and a unique ray k in $\overrightarrow{ABr'}$ with $\overrightarrow{ABk} = z$ (Figure 12.4).

Corollary 11.11 implies that $\overrightarrow{ABr} \cup \overrightarrow{ABr'}$ comprise all the rays in $\boldsymbol{P}_A$. So h and k are the only two rays that form with $\overrightarrow{AB}$ an angle of measure z. (Note that what we have just proved constitutes Problem 8 in Chapter 11.)

The interior points of r and r' are in opposite halfplanes with edge $\overleftrightarrow{AB}$ (Theorem 10.3). Hence, so are h^o and k^o, by Theorem 12.2.

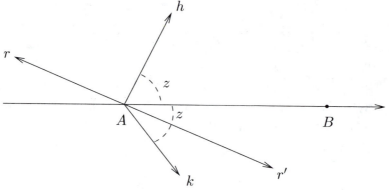

Figure 12.4

Theorem 12.4. (The Crossbar Theorem) If $\underline{hk}$ is a proper angle with vertex X, if ray j lies between h and k, and if A and C are interior points of h and k, respectively, then there is an interior point B of j with A-B-C (Figure 12.5).

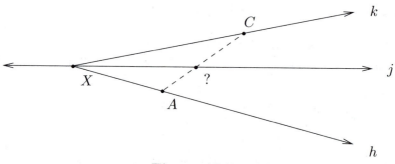

Figure 12.5

Proof. Note that $\underline{hk}$ a proper angle implies that $C \notin \overleftrightarrow{XA}$. Hence, $0 < AC < \omega$ by Theorem 10.5. Also, h-j-k forces $\underline{hj}$ and $\underline{jk}$ to be proper angles. It follows that neither A nor C is on the line $j \cup j'$. Then by a choice of reasons, X, A, and C are noncollinear.

Suppose that A and C are on the same side of $j \cup j'$. Then k must be in the fan $\overrightarrow{jh}$ (Theorem 12.2). Hence either j-h-k or j-k-h, which contradicts h-j-k and Theorem 11.3.

So A and C are on opposite sides of $j \cup j'$. Then Axiom S forces $\overline{AC}$ to meet either j or j'.

Suppose that j' meets $\overline{AC}$. Then there is a point Y on j' with A-Y-C. So Axiom C implies that h-j'-k. Then by Proposition 11.7, $180 = jj' \le hk$, a contradiction. Therefore, j meets $\overline{AC}$, but not at X (or X^*). The theorem is proved.

The Crossbar Theorem is often applied in the following context: Let $\triangle XAC$ be a triangle, and assume that j is any ray with endpoint X such that j is between $\overrightarrow{XA}$ and $\overrightarrow{XC}$ (as in Figure 12.5). Then the Crossbar Theorem guarantees that j meets $\overline{AC}$, the side of the triangle that is opposite vertex X. This fact seems quite obvious in the Euclidean plane. Indeed, Euclid employed it without any explicit justification.

Now we introduce a more specific notation for halfplanes. For any line m and point A not on m,

$$H(A, m) = \text{the halfplane with edge } m \text{ that contains } A.$$

Definition. Let $\angle AXC$ be a proper (nondegenerate, nonstraight) angle. The *interior* of $\angle AXC$, written Int $\angle AXC$, is the set

$$\text{Int } \angle AXC = H(A, \overleftrightarrow{XC}) \cap H(C, \overleftrightarrow{XA})$$

(See Figure 12.6.)

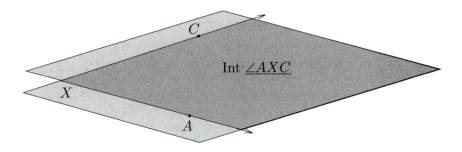

Figure 12.6

Examples

Int $\angle \underline{AXC}$ in $\mathbb{H}$ (Figure 12.7)

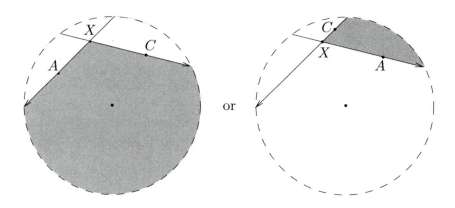

or

Figure 12.7

Int $\angle \underline{AXC}$ in $\mathbb{S}$ (Figure 12.8)

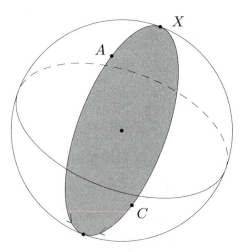

Figure 12.8

Next we show that the interior of an angle depends only on the two rays that define the angle and not on the choice of interior points on the rays (A and C) that are used in the definition of the interior.

Proposition 12.5. Let $\angle AXC$ be a proper angle. If P is any interior point of $\overrightarrow{XA}$ and Q is any interior point of $\overrightarrow{XC}$, then Int $\angle AXC =$ Int $\angle PXQ$ (Figure 12.9).

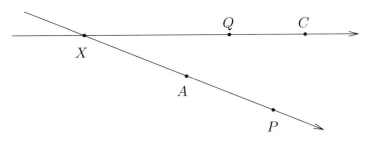

Figure 12.9

Proof. Problem 2.

Proposition 12.6. The interior of any proper angle is a convex set.

Proof. Problem 3.

Theorem 12.7. For any proper angle $\angle AXC$,

$$\text{Int } \angle AXC \ = \ \bigcup_{j} j^{o},$$

where the union is over all rays j such that $\overrightarrow{XA}\text{-}j\text{-}\overrightarrow{XC}$.

Proof. Problem 4.

Problem Set 12

1. Show by a picture that the Crossbar Theorem is *false* in $\mathbb{G}$.

2. Prove Proposition 12.5.

3. Prove Proposition 12.6.

4. Prove Theorem 12.7.

5. Prove the following: Let $\angle AXC$ be a proper angle. If P is an interior point of $\overrightarrow{XA}$ and Q is an interior point of $\overrightarrow{XC}$, then every point B with $P\text{-}B\text{-}Q$ lies in Int $\angle AXC$ (Figure 12.10).

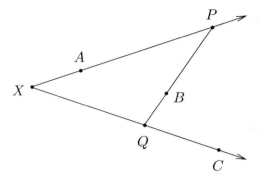

Figure 12.10

6. Is the converse of Problem 5 true? Must every point B of Int $\angle AXC$ lie on some segment $\overline{PQ}$ with P an interior point of $\overrightarrow{XA}$ and Q an interior point of $\overrightarrow{XC}$? (Hint: Look at $\mathbb{H}$.)

7. Assume that $AC < \omega$, $A\text{-}B\text{-}C$, and D is a point not on $\overleftrightarrow{AC}$. Let h be a ray with endpoint C such that h meets $\overline{BD}^o$. Prove that h meets $\overline{AD}$.

8. Suppose that A, P and R are noncollinear, $A\text{-}X\text{-}P$, $A\text{-}Z\text{-}R$, and $P\text{-}Q\text{-}R$.

   (a) Prove there is a point Y on $\overrightarrow{AQ}$ so that $X\text{-}Y\text{-}Z$.

   (b) Prove further that $A\text{-}Y\text{-}Q$.

9. Let A, B, C be three noncollinear points. The definition of the *triangle* $\triangle ABC$ is given in the next section as $\overline{AB} \cup \overline{BC} \cup \overline{CA}$. Define the *interior of a triangle* by

$$\text{Int}\,(\triangle ABC) = H(A, \overleftrightarrow{BC}) \cap H(B, \overleftrightarrow{AC}) \cap H(C, \overleftrightarrow{AB}).$$

   (a) Prove that Int $(\triangle ABC)$ = Int $\angle ABC \cap$ Int $\angle BCA$.

   (b) If $\triangle ABC$ is a triangle and if h, j are rays with $\overrightarrow{AB}\text{-}h\text{-}\overrightarrow{AC}$ and $\overrightarrow{BA}\text{-}j\text{-}\overrightarrow{BC}$, prove that h and j meet in a point of Int $(\triangle ABC)$.

10. Suppose that A, B, C are three noncollinear points, and $B\text{-}X\text{-}C$. For any real number z with $0 \leq z \leq 180$, prove that there is a ray j with endpoint X so that $\overrightarrow{XC}j = z$ and j meets either $\overline{AB}$ or $\overline{AC}$.

11. Let B and C be distinct points on the same side of line $\overleftrightarrow{AX}$. Prove that one and only one of the following holds: A-B-C, A-C-B, $\overrightarrow{AX}$-$\overrightarrow{AB}$-$\overrightarrow{AC}$, or $\overrightarrow{AX}$-$\overrightarrow{AC}$-$\overrightarrow{AB}$.

12. Assume that $\omega < \infty$ and that rays $\overrightarrow{AB}$, $\overrightarrow{AC}$, $\overrightarrow{AD}$ are such that $\angle BAD$ is proper, $\overrightarrow{AB}$-$\overrightarrow{AC}$-$\overrightarrow{AD}$, and $\angle BAC > \angle DAC$. For each statement, answer true or false and explain why.

   (a) $\overrightarrow{AC} \cap \overline{BD} = \emptyset$.

   (b) C and D are in the same halfplane with edge $\overleftrightarrow{AB}$.

   (c) C and C^* are in the same halfplane with edge $\overleftrightarrow{AB}$.

   (d) C and B^* are in the same halfplane with edge $\overleftrightarrow{AB}$.

   (e) $\angle DAC < 90$.

   (f) There is another ray $\overrightarrow{AE}$ so that $\angle EAC = \angle DAC$ and D and E are in the same halfplane with edge $\overleftrightarrow{AB}$.

   (g) $\overrightarrow{AB^*}$-$\overrightarrow{AC}$-$\overrightarrow{AD}$.

13   Side-Angle-Side

In this chapter we define "congruence," first for segments and angles and then for triangles (once we have defined what they are). Then we add an axiom (the last one!) to our abstract geometry to govern the properties of congruence.

Definition.   Two segments are called *congruent* if they have the same length; so using the notation "$\cong$" to denote "congruent," we have

$$\overline{AB} \cong \overline{CD} \;\Leftrightarrow\; AB = CD.$$

Two angles are called *congruent* if they have the same measure; so

$$\angle ABC \cong \angle XYZ \;\Leftrightarrow\; \angle ABC = \angle XYZ.$$

Before we define "triangle," we recall that if A, B, and C are three noncollinear points, then the distances AB, BC, and AC are all less than ω (as otherwise, if, say, $AB = \omega$, then B is on $\overleftrightarrow{AC}$ by Theorem 10.5). Thus segments $\overline{AB}$, $\overline{BC}$, and $\overline{AC}$ are defined.

Definition.   Let A, B, and C be three noncollinear points (Figure 13.1). The *triangle* $\triangle ABC$ with vertices A, B, C is defined as the set

$$\triangle ABC = \overline{AB} \cup \overline{BC} \cup \overline{CA}.$$

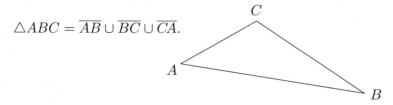

Figure 13.1

Segments $\overline{AB}, \overline{BC}, \overline{CA}$ are the *sides* of $\triangle ABC$; lines $\overleftrightarrow{AB}, \overleftrightarrow{BC}, \overleftrightarrow{CA}$ are the *sidelines* of $\triangle ABC$; and angles $\angle CAB$ ($\angle A$), $\angle ABC$ ($\angle B$), $\angle BCA$ ($\angle C$) are the *angles* of $\triangle ABC$. $\angle CAB$ and vertex A are called *opposite* side $\overline{BC}$, and the term is defined similarly for the other angles, vertices, and sides. As before, $\angle A$ is said to be *included* between sides $\overline{AB}$ and $\overline{AC}$.

The sum $\angle A + \angle B + \angle C$ of the measures of the angles of $\triangle ABC$ is called the *angle sum* of $\triangle ABC$, denoted $\sigma(ABC)$. Be forewarned that $\sigma(ABC)$ is *not* always 180 in our general system.

Definition. A correspondence $A \leftrightarrow X, B \leftrightarrow Y,$ and $C \leftrightarrow Z$ (abbreviated $ABC \leftrightarrow XYZ$) between the vertices of $\triangle ABC$ and those of $\triangle XYZ$ is called a *congruence* if all corresponding sides and angles are congruent; that is,

$$\overline{AB} \cong \overline{XY}, \qquad \overline{BC} \cong \overline{YZ}, \qquad \overline{CA} \cong \overline{ZX},$$
$$\angle ABC \cong \angle XYZ, \qquad \angle BCA \cong \angle YZX, \qquad \angle CAB \cong \angle ZXY.$$

We denote this by $\triangle ABC \cong \triangle XYZ$ and say that $\triangle ABC$ is *congruent* to $\triangle XYZ$ *under the correspondence* $ABC \leftrightarrow XYZ$.

Note: A congruence $\triangle ABC \cong \triangle XYZ$ always depends on a *specific* correspondence of vertices and may not remain valid if the vertices of either triangle are listed in a different order. For example, consider a triangle in $\mathbb{E}$ as in Figure 13.2.

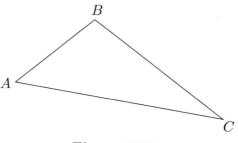

Figure 13.2

Here, $\triangle ABC = \triangle ACB$ (both equal the same set $\overline{AB} \cup \overline{BC} \cup \overline{CA}$), but $\triangle ABC \not\cong \triangle ACB$. That is, $ABC \leftrightarrow ACB$ is *not* a congruence, as $AB \neq AC$ implies that $\overline{AB} \not\cong \overline{AC}$. On the other hand, we always have $\triangle ABC \cong \triangle ABC$.

We now present the final basic assumption of our abstract system. An example and a discussion later in the section help to show why this axiom is necessary.

Congruence Axiom (Axiom SAS [Side-Angle-Side]):

If, under the correspondence $ABC \leftrightarrow XYZ$ between the vertices of $\triangle ABC$ and those of $\triangle XYZ$, two sides and the included angle of the first triangle are congruent, respectively, to the corresponding two sides and included angle of the second triangle, then $\triangle ABC \cong \triangle XYZ$ (Figure 13.3).

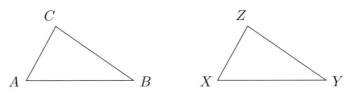

Figure 13.3

Axiom SAS says, for example, that if $\overline{AC} \cong \overline{XZ}$, $\overline{AB} \cong \overline{XY}$ and $\angle CAB \cong \angle ZXY$, then $\overline{BC} \cong \overline{YZ}$, $\angle ABC \cong \angle XYZ$ and $\angle ACB \cong \angle XZY$.

The next example shows that Axiom SAS cannot be proved as a consequence of the first 20 axioms.

Example. $\mathbb{M}$ satisfies all previous axioms, but *not* SAS: Consider the points $A(0,0)$, $B(2,0)$, $C(0,2)$, $X(3,0)$, $Y(4,1)$, and $Z(2,1)$ in $\mathbb{M}$, and the correspondence $ABC \leftrightarrow XYZ$ (Figure 13.4).

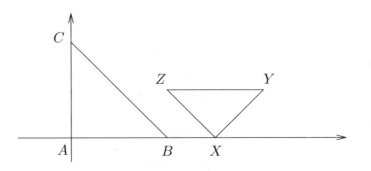

Figure 13.4

Now $AB = 2 = XY$, $AC = 2 = XZ$ ($\mathbb{M}$ distance), and $\angle CAB = 90 = \angle ZXY$. So two sides $\overline{AB}$, $\overline{AC}$, and the included angle $\angle CAB$ of $\triangle ABC$ are congruent, respectively, to the corresponding two sides $\overline{XY}$, $\overline{XZ}$, and included angle $\angle ZXY$ of $\triangle XYZ$.

But the correspondence $ABC \leftrightarrow XYZ$ is *not* a congruence, since $BC = 4 \neq 2 = YZ$ and hence $\overline{BC} \not\cong \overline{YZ}$. So $\triangle ABC \not\cong \triangle XYZ$.

We will accept without proof the (fairly obvious) statement that $\mathbb{E}$ and $\mathbb{S}$ satisfy Axiom SAS and the (not at all obvious) statement that $\mathbb{H}$ satisfies SAS. This fact is proved in Appendix II.

Definition. The axiom system with undefined terms "point," "line," "distance," and "angle measure," and axioms D1–D3, I1–I4, BP, QP, N, RR, S, M1–M4, BR, QR, RF, C, and SAS is called *absolute*, or *neutral, plane geometry*. Any model of the system is called an *absolute plane*.

$\mathbb{E}$, $\mathbb{S}$, and $\mathbb{H}$ are examples of absolute planes.

Here we state again, for the convenience of the reader, the 21 axioms that define an absolute plane:

D1.   (Positivity) $PQ \geq 0$, for all points P and Q.

D2.   (Definiteness) $PQ = 0$ if and only if $P = Q$, for all points P and Q.

D3.   (Symmetry) $PQ = QP$, for all points P and Q.

I1.   There are at least two different lines.

I2.   Each line contains at least two different points.

I3.   Each two different points lie in *at least* one line.

I4.   Each two different points P, Q, *with $PQ < \omega$*, lie in *at most* one line.

BP.   If A, B, and C are different, collinear points and if $AB + BC \leq \omega$, then there exists a betweenness relation among A, B, and C.

QP.   If A, B, C, X are distinct, collinear points, and if A-B-C, then at least one of the following must hold:

$$X\text{-}A\text{-}B, \quad A\text{-}X\text{-}B, \quad B\text{-}X\text{-}C, \text{ or } B\text{-}C\text{-}X.$$

N.   For any point A on a line l, there exists a point B on l with $0 < AB < \omega$.

RR.   For any ray $\overrightarrow{AB}$ and any real number s with $0 \leq s \leq \omega$ ($s < \omega$ when $\omega = \infty$), there is a point X in $\overrightarrow{AB}$ with $AX = s$.

S.   For each line m, there exists a pair of opposite halfplanes with edge m.

M1.   (Positivity) $0 \leq pq \leq 180$, for all coterminal rays p and q.

M2.   (Definiteness) $pq = 0$ if and only if $p = q$, for all coterminal rays p and q.

M3.   (Symmetry) $pq = qp$, for all coterminal rays p and q.

M4.   (Opposites) $pq = 180$ if and only if $q = p'$, for all coterminal rays p and q.

BR.   If a, b, and c are different, coterminal rays and if $ab + bc \leq 180$, then there exists a betweenness relation among a, b, and c.

QR.   If a, b, c, x are distinct coterminal rays and $a\text{-}b\text{-}c$, then at least one of the following must hold: $x\text{-}a\text{-}b$, $a\text{-}x\text{-}b$, $b\text{-}x\text{-}c$, or $b\text{-}c\text{-}x$.

RF.   For any fan $\overrightarrow{pq}$ and any real number t with $0 \leq t \leq 180$, there is a ray r in $\overrightarrow{pq}$ with $pr = t$.

C.   Let A, B, and C be points on a line m, let X be a point not on m, and let $a = \overrightarrow{XA}$, $b = \overrightarrow{XB}$, and $c = \overrightarrow{XC}$. If $A\text{-}B\text{-}C$, then $a\text{-}b\text{-}c$.

SAS.   If, under the correspondence $ABC \leftrightarrow XYZ$ between the vertices of $\triangle ABC$ and those of $\triangle XYZ$, two sides and the included angle of the first triangle are congruent, respectively, to the corresponding two sides and included angle of the second triangle, then $\triangle ABC \cong \triangle XYZ$.

Now that the introduction of axioms is complete, there are some questions that it seems natural to consider:

(1) Are all 21 axioms really necessary, or could a smaller number lead us to the same place where we are now?

(2) If we only wish to do Euclidean geometry, what is wrong with simply following Euclid's definitions and axioms (postulates) (as listed in Appendix I)?

There seems to be no redundancy among the 21 axioms we have given. (Find some, and your instructor will give you a lot of extra credit!) There are, however, alternative formulations that use fewer axioms and yet arrive at the same system. The problem with such alternatives for our system of absolute geometry is that some of the earlier proofs would be more complicated than those we have done.

We quickly present the axioms of one alternative scheme, just for comparison. For simplicity, we adopt the hypothesis, implicit in the next axiom, that $\omega = \infty$ (so models for the system include $\mathbb{E}$ and $\mathbb{H}$, but not $\mathbb{S}$). This simplifying assumption allows portions of the theory of absolute planes to be developed in a more streamlined manner than seems possible for our more general context. The terms and notation for points, lines, distance, angles and angle measure are as before.

Ruler Axiom (Axiom R):

For any line m, there is a function $f : m \to \mathbb{R}$ that is one to one and onto $\mathbb{R}$ and such that for all points X, Y on m,

$$XY = |f(X) - f(Y)|.$$

Protractor Axiom (Axiom P):

For each pencil $\boldsymbol{P}$ of rays, there is a function $g : \boldsymbol{P} \to (-180, 180]$ that is one to one and onto $(-180, 180]$, and such that for any rays h, k in $\boldsymbol{P}$,

$$hk = \begin{cases} |g(h) - g(k)| & \text{if } |g(h) - g(k)| \leq 180 \\ 360 - |g(h) - g(k)| & \text{if } |g(h) - g(k)| > 180. \end{cases}$$

Now the following list of axioms yields a system that is equivalent to absolute plane geometry for "unbounded planes" ($\omega = \infty$). That is, each of our 21 axioms (with $\omega = \infty$) can be proved as a theorem given the following statements, and each of the statements can be proved as theorems from our 21 axioms (with $\omega = \infty$).

Axiom I1 (as before);

Axiom I3′ (new): Each two different points lie in exactly one line;

Axiom R;

Axiom P;

Axiom M4 (as before);

Axioms C, S and SAS (as before);

a total of eight axioms.

There may be a completely different (and as yet unknown) approach that would yield a more efficient development of the foundations of geometry. We could follow the analytic method for the Euclidean plane; that is, define *points* as ordered pairs (x, y) of real numbers, and define *distance* by the distance formula and *lines* by the usual equations. Then all geometric theorems may be proved by algebra and/or calculus. This approach also shows that the *existence* (or the logical consistency) of the Euclidean plane follows from the existence of the real numbers. But it does not fall within the bounds of the conceptual game we are playing.

The axiom system studied in this course is a variation on work of G. D. Birkhoff (1932), who introduced the Ruler and Protractor Axioms. His results on the Euclidean plane were preceded by several rigorous (by modern standards) axiomatic treatments of Euclidean geometry. These include the work of Moritz Pasch (1882), Giuseppe Peano (1889), and David Hilbert (1902). Hilbert's axioms make no reference to the real numbers, but they are more complicated than Birkhoff's.

Now we make a few remarks about the question of following Euclid's exposition. First, Euclid has a number of definitions (including "point," "line," "straight line," "surface") that seem obscure and are ultimately irrelevant. The plane seems to have been a concrete, real object to the Greek geometers, and not merely an abstract, logical concept. Euclid's definitions were attempts to describe precisely certain features of this object. But in proving subsequent theorems about geometry, these "definitions" were never used.

Second, there are properties of the plane (involving betweenness and separation) that are tacitly invoked by Euclid without any explicit recognition of the underlying concepts. Our Axioms C and S (and consequences such as the Crossbar Theorem) set forth these concepts and make clear exactly what is to be assumed about them.

Third, the Side-Angle-Side criterion for congruence of triangles is "proved" by Euclid by an argument that essentially says "place one triangle on top of the other" (the principle of superposition). But there is nothing in his postulates that permits such an operation. It is an intuitively obvious property of his plane. But our example $\mathbb{M}$ shows that Axiom SAS does not logically follow from the other axioms. So to *assume* it as another axiom seems necessary and fills another gap in Euclid's work.

We must note that the flaws in the original foundations of Euclidean geometry are very minor when compared with the fantastic intellectual achievements of the overall work. Once we correct (or can safely ignore) the gaps in Euclid's system, then the development of Euclidean geometry proceeds smoothly.

It is an interesting exercise to compare Euclid's postulates with the axioms and theorems of this course. We return now to general absolute geometry and prove more consequences of all of our axioms (including SAS). The first one is the familiar Angle-Side-Angle criterion for congruence of triangles.

Theorem 13.1. (ASA) If, under the correspondence $ABC \leftrightarrow XYZ$ between the vertices of $\triangle ABC$ and those of $\triangle XYZ$, two angles and the included side of the first triangle are congruent, respectively, to the corresponding two angles and included side of the second, then $\triangle ABC \cong \triangle XYZ$.

Proof. We may assume that $\angle A = \angle X$, $\angle B = \angle Y$, and $AB = XY$ (Figure 13.5).

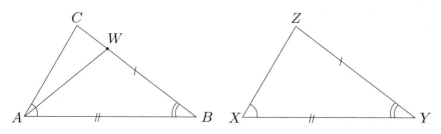

Figure 13.5

Suppose (toward a contradiction) that $BC \neq YZ$. Then we may assume $BC > YZ$ (otherwise, just switch notation for the roles of the two triangles: $B \leftrightarrow Y$, etc.).

The Unique Distances Theorem (Theorem 8.6) implies that there is a point W on $\overleftrightarrow{BC}$ with B-W-C and $BW = YZ$. $\triangle ABW$ exists (as $BW = YZ < \omega$ means that line $\overleftrightarrow{BW}$ is defined; then $\overleftrightarrow{BW} = \overleftrightarrow{BC} \neq \overleftrightarrow{AB}$, so A, B and W are not collinear). $\overrightarrow{BW} = \overrightarrow{BC}$ by Theorem 8.4, so $\angle ABC = \angle ABW$.

Now $AB = XY$, $BW = YZ$ and $\angle ABW = \angle XYZ$. So Axiom SAS implies that $\triangle ABW \cong \triangle XYZ$. In particular, $\angle BAW = \angle YXZ = \angle X$, which equals $\angle BAC$ by hypothesis.

But B-W-C and Axiom C yield $\overrightarrow{AB}$-$\overrightarrow{AW}$-$\overrightarrow{AC}$. Then by definition of betweenness of rays, $\angle BAW + \angle WAC = \angle BAC$. So

$$\angle BAW = \angle BAC = \angle BAW + \angle WAC,$$

hence $\angle WAC = 0$. Then $\overrightarrow{AW} = \overrightarrow{AC}$ by Axiom M2, which contradicts $\overrightarrow{AB}$-$\overrightarrow{AW}$-$\overrightarrow{AC}$.

It follows that $BC = YZ$. Now $AB = XY$, $BC = YZ$, and $\angle B = \angle Y$. So Axiom SAS implies that $\triangle ABC \cong \triangle XYZ$.

Definition. $\triangle ABC$ is called *isosceles* if two sides have the same length; *equilateral* if all three sides have the same length; *scalene* if all three sides have different lengths; and *equiangular* if all three angles have the same measure.

The classical Latin name for the next theorem means "asses' bridge"; it came from a figure in Euclid's proof.

Theorem 13.2. (*Pons asinorum*) In any $\triangle ABC$, $AB = AC$ if and only if $\angle ABC = \angle ACB$ (Figure 13.6).

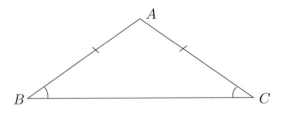

Figure 13.6

Proof. Consider the correspondence $ABC \leftrightarrow ACB$, which sends the vertices of $\triangle ABC$ to themselves, in the order $A \leftrightarrow A$, $B \leftrightarrow C$,

$C \leftrightarrow B$. Recall that we always have $BC = CB$ (by Axiom D3) and $\angle BAC = \angle CAB$ (by Axiom M3).

If $AB = AC$, then the three congruences $\overline{AB} \cong \overline{AC}$, $\overline{AC} \cong \overline{AB}$, and $\underline{\angle BAC} \cong \underline{\angle CAB}$ imply by Axiom SAS that $\triangle ABC \cong \triangle ACB$. Hence, $\angle ABC = \angle ACB$ by definition of congruence of triangles.

Conversely, if $\angle ABC = \angle ACB$ then the congruences $\angle ABC \cong \angle ACB$, $\angle ACB \cong \angle ABC$, and $\overline{BC} \cong \overline{CB}$ imply by Theorem 13.1 that $\triangle ABC \cong \triangle ACB$. Hence, $AB = AC$.

Corollary 13.3. A triangle is equilateral if and only if it is equiangular.

Proof. Problem 4.

The final result in this chapter contains the third criterion for congruence of triangles which is valid in all absolute planes: the Side-Side-Side Theorem.

Theorem 13.4. (SSS) If in $\triangle ABC$ and $\triangle XYZ$, $AB = XY$, $BC = YZ$ and $CA = ZX$, then $\triangle ABC \cong \triangle XYZ$.

Proof. Given the hypotheses, our goal is to show that $\angle A = \angle X$. For then the desired conclusion will follow from Axiom SAS.

Suppose (toward a contradiction) that $\angle A \neq \angle X$. By interchanging notation for the two triangles if necessary, we may assume that $\angle A > \angle X$. The Unique Angular Distances for Fans Theorem (Theorem 11.6), applied to $\overrightarrow{AB}\overrightarrow{AC}$, yields that there is a ray h with endpoint A so that $\overrightarrow{AB}\text{-}h\text{-}\overrightarrow{AC}$ and $\overrightarrow{AB}h = \angle X$. By the Crossbar Theorem (Theorem 12.4) there is a point $D \in h^{\circ}$ (so $h = \overrightarrow{AD}$) with $B\text{-}D\text{-}C$. Thus $\angle BAD = \angle X$ (Figure 13.7).

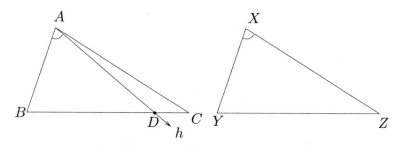

Figure 13.7

There is a point $P \in \overrightarrow{AD}$ with $AP = XZ$ (Axiom RR). Then $\triangle ABP \cong \triangle XYZ$ (Axiom SAS). By hypothesis and the definition of congruent triangles, $BC = YZ = BP$. Since B-D-C implies that $BD < BC$, it follows that $P \neq D$. So either A-P-D or A-D-P (Figure 13.8).

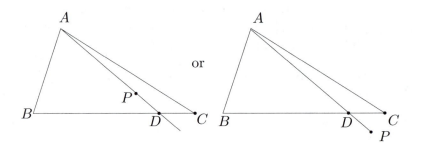

Figure 13.8

Now $BP = BC$ implies that $\angle BCP = \angle BPC$ (*pons asinorum* (Theorem 13.2)). The hypotheses and $\triangle ABP \cong \triangle XYZ$ yield that $AC = XZ = AP$. So $\angle APC = \angle ACP$, again by Theorem 13.2. Axiom C and B-D-C give us that $\angle BPD + \angle DPC = \angle BPC$, hence $\angle BPC > \angle DPC$.

Suppose that A-P-D. Then $\overrightarrow{PD} = \overrightarrow{PA}'$ (Theorem 9.6), and Theorem 11.8 and Axiom M4 yield that $\angle APC + \angle DPC = 180$. Also, $\angle ACP + \angle BCP = \angle ACB$ (A-P-D and Axiom C). Since $\angle APC = \angle ACP$ and $\angle BCP = \angle BPC > \angle DPC$, $\angle ACB = \angle ACP + \angle BCP > \angle APC + \angle DPC = 180$, which contradicts Axiom M1.

In the case where A-D-P, a similar argument leads to another contradiction. This is left as an exercise (Problem 12).

Problem Set 13

1. To each of Euclid's postulates 1–4, find which one (or more) of our axioms and/or theorems is/are most similar in content. Explain briefly in each case.

2. If M is the midpoint of $\overline{AB}$ and $\angle BMD = 90$, show that $AD = BD$.

3. Suppose that A, C, and E are noncollinear, A-B-C, A-D-E, $AC = AE$, and $BC = DE$. Prove that $CD = BE$.

4. Prove Corollary 13.3.

5. If $\angle AMD = 90$, if A and P lie on opposite sides of $\overleftrightarrow{DM}$, and if $\angle PDM = \angle ADM$, show that the rays $\overrightarrow{DP}$ and $\overrightarrow{MA}'$ (opposite) intersect. (Hints: (a) There exists a point E in $\overrightarrow{MA}'$ such that $ME = MA$ (why and so what?) (b) Corollary 12.3.)

6. Suppose that A, B, C, D are collinear, that no two of these points are collinear with X, that $\overrightarrow{XA}$-$\overrightarrow{XB}$-$\overrightarrow{XC}$-$\overrightarrow{XD}$, $\angle XAB = \angle XDC$, and $\angle AXC = \angle BXD$. Show that $XB = XC$.

7. Is ASA a valid congruence criterion in $\mathbb{M}$?

8. Suppose that A, B, C are three noncollinear points. Prove that there exists a fourth point D not on $\overleftrightarrow{AB}$ so that $\triangle ABC \cong \triangle ABD$.

9. Suppose that $\omega < \infty$, that P, Q, R are noncollinear points with $P^* =$ antipode of P, and that $\angle PQR = 30$ and $\angle PRQ = 150$. Prove that $\triangle P^*QR \cong \triangle PRQ$.

10. Suppose that $\omega < \infty$ and A, B, C are noncollinear points. Show that every point in $\mathbb{P}$ not on any of the lines $\overleftrightarrow{AB}$, $\overleftrightarrow{BC}$, $\overleftrightarrow{CA}$ lies in the interior (Problem 9, Chapter 12) of exactly one of the eight triangles $\triangle ABC$, $\triangle A^*BC$, $\triangle AB^*C$, $\triangle ABC^*$, $\triangle AB^*C^*$, $\triangle A^*BC^*$, $\triangle A^*B^*C$, and $\triangle A^*B^*C^*$. (So a plane of finite diameter can be regarded as a union of exactly eight nonoverlapping [except for sides] triangles.)

11. Suppose that $\omega < \infty$ and $\triangle ABC$ is trirectangular (that is, each of $\angle A$, $\angle B$, $\angle C$ is a right angle). Show that each of $\triangle A^*BC$, $\triangle AB^*C$, $\triangle ABC^*$, $\triangle AB^*C^*$, $\triangle A^*BC^*$, $\triangle A^*B^*C$, and $\triangle A^*B^*C^*$ is trirectangular.

12. Complete the proof of Theorem 13.4.

13. Prove that the statement of the Ruler Axiom is a *theorem* in absolute geometry with $\omega = \infty$. (Hint: For any line m,

let A be a point on m and let h, h' be a pair of opposite
rays with endpoint A whose union is m (from the Opposite
Ray Theorem).   For any point X on m , define the function
$f : m \to \mathbb{R}$ by

$$f(X) = \begin{cases} AX & \text{if } X \in h \\ -AX & \text{if } X \in h' \end{cases}$$

Prove that f has all of the properties listed in the Ruler
Axiom.)

14. Prove that the statement of the Protractor Axiom is a *theorem*
in absolute geometry. (Hint: For any pencil $\boldsymbol{P}_A$, let r and
r' be a pair of opposite rays with endpoint A whose union is
a line m (via the Opposite Ray Theorem). Let H and K
be the halfplanes with edge m . If j is any ray in $\boldsymbol{P}_A$, let
j^o denote the set of interior points of j . Then for $j \neq r$ or
r' , either $j^o \subseteq H$ or $j^o \subseteq K$ by Theorem 10.3. Define the
function $g : \boldsymbol{P}_A \to (-180, 180]$ by, for all $j \in \boldsymbol{P}_A$,

$$g(j) = \begin{cases} rj & \text{if } j^o \subseteq H \\ -rj & \text{if } j^o \subseteq K \\ 0 & \text{if } j = r \\ 180 & \text{if } j = r' \end{cases}$$

Prove that g has all of the properties listed in the Protractor
Axiom.)

15. Assume that $\omega < \infty$ and that A, B, C are three noncollinear
points. Prove that $\triangle ABC \cong \triangle A^*B^*C^*$.

14 Perpendiculars

We continue to develop absolute plane geometry by examining the notion of perpendicular lines. For planes of infinite diameter, the results of our theory resemble those of high school geometry. But when ω is finite some striking differences emerge. There can be more than one perpendicular from a given point to a given line, under certain conditions.

Definition. Two angles are *supplementary* if the sum of their measures is 180 and *complementary* if the sum of their measures is 90.

Recall from Chapter 11 that a proper angle is *acute, right,* or *obtuse* as its measure is respectively less than, equal to, or greater than 90.

Definition. Two angles $\underline{hk}$ and $\underline{rs}$ are *vertical* if $\{r, s\} = \{h', k'\}$ (opposite rays).

Although we include proofs here, the next two theorems may be justified as the duals of Problems 1 and 2 of Chapter 10.

Theorem 14.1. (Supplementary Angles Theorem) If h, j are coterminal and h' is opposite h, then $\underline{hj}$ and $\underline{jh'}$ are supplementary (Figure 14.1).

Figure 14.1

Proof. If $j = h$, then $hj = 0$ and $jh' = 180$ by Axioms M2 and M4. If $j = h'$, then $hj = 180$ and $jh' = 0$. So in these cases $hj + jh' = 180$. If $j \neq h$ or h', then h-j-h' by Theorem 11.8. Thus $hj + jh = hh'$, while $hh' = 180$ by Axiom M4.

Theorem 14.2. Vertical angles are congruent (Figure 14.2).

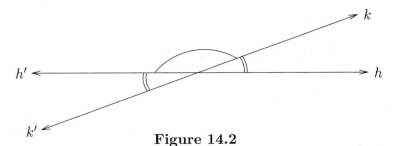

Figure 14.2

Proof. Since $h'k' = k'h'$ by Axiom M3, it suffices to show $h'k' = hk$. If $k = h$, then $k' = h'$ and $h'k' = hk = 0$ by Axiom M2. If $k = h'$, then $h'k' = kk' = 180 = hh' = hk$ by Axiom M4.

If $k \neq h$ or h', then h-k-h' and k-h'-k' by Theorem 11.8. So

$$hk + kh' = hh' = 180 = kk' = kh' + h'k'.$$

Thus $hk + kh' = h'k' + kh'$, from which it follows that $hk = h'k'$.

Remark. When two lines meet, four coterminal rays are involved, of the form h, h', k, k', as in Figure 14.1. If $hk = 90$, then $h'k' = 90$ by Theorem 14.2 and $kh' = 90 = hk'$ by the Supplementary Angles Theorem. So if one of the four angles $\underline{hk}, \underline{hk'}, \underline{h'k}, \underline{h'k'}$ formed by two intersecting lines is a right angle, then all four are right angles (Figure 14.3).

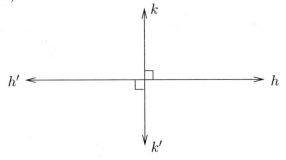

Figure 14.3

Definition. Two intersecting lines are *perpendicular* (at point of intersection B) if the four angles they determine (with vertex B) are right angles. When lines m and n are perpendicular, write $m \perp n$.

If $\omega < \infty$ and lines m and n meet in a point B, then they also meet in B^*, by Theorem 10.8. It follows from Proposition 11.14(a) that if $m \perp n$ (at B), then $m \perp n$ (at B^*) as well, so there is nothing ambiguous about perpendicularity in this sense.

Theorem 14.3. Through a point A on a line m there is exactly one line n perpendicular to m (Figure 14.4).

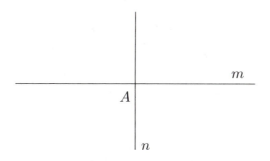

Figure 14.4

Proof. Let h be a ray on m with endpoint A. Corollary 12.3 implies that there is a ray k in $\boldsymbol{P}_A$ with $hk = 90$. Let n be the carrier of k. Then $n = k \cup k'$, and k and k' are the *only* rays in $\boldsymbol{P}_A$ that form with h an angle of measure 90, by the Supplementary Angles Theorem and Corollary 12.3. So n is the unique perpendicular to m at A.

The next three theorems deal with the existence of perpendiculars to a given line through a given point not on the line. There is always at least one such perpendicular (Theorem 14.4); and in just one situation, which we describe precisely, there are multiple such perpendiculars. Our model in the latter case is $\mathbb{S}$, where *any* line (i.e., great circle) through the north pole is perpendicular to the equator.

Theorem 14.4. Through a point A not on a given line m there is at least one line n perpendicular to m.

Proof. Choose any point B on m. A not on m implies that $AB < \omega$ (by Theorem 10.8). If $\overleftrightarrow{AB} \perp m$, then we are done, so we may assume that $\overleftrightarrow{AB}$ is not perpendicular to m. Let h, h' be the two opposite rays with endpoint B such that $m = h \cup h'$. Let $k = \overrightarrow{BA}$ (Figure 14.5).

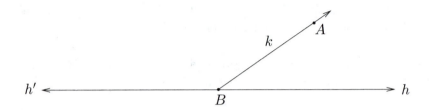

Figure 14.5

Since $hk \neq 90$, we have that either $hk < 90$ or $h'k < 90$ by the Supplementary Angles Theorem. We may assume $hk < 90$. By Theorem 11.16, there is a ray $j \in \boldsymbol{P}_B$ with k-h-j and $kh = hj$. Now

$$0 < kj = kh + hj = 2kh = 2hk < 180,$$

so kj is a proper angle. By Axiom RR, there is a point C on j with $BC = BA$. The Crossbar Theorem implies that there is a point X on h with A-X-C (Figure 14.6).

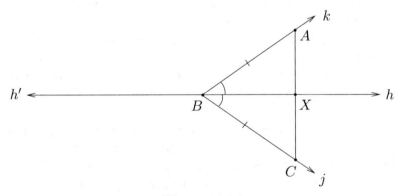

Figure 14.6

A, B, X are noncollinear as $k \neq h$ or h'; C, B, X are noncollinear as $j \neq h$ or h'. So triangles $\triangle ABX$, $\triangle CBX$ are defined.

Now $BX = BX$, $BC = BA$ and $\angle CBX = jh = hj = kh = \angle ABX$. Hence $\triangle ABX \cong \triangle CBX$ by Axiom SAS. So, in particular,

$\angle AXB = \angle CXB$. Now A-X-C implies that $\overrightarrow{XC} = \overrightarrow{XA}'$ by the Opposite Ray Theorem. So by Theorem 14.1, angles $\overrightarrow{XAXB} = \angle AXB$ and $\overrightarrow{XCXB} = \angle CXB$ are supplementary. Thus $\angle AXB + \angle CXB = 180$. Hence $\angle AXB = \angle CXB = 90$, and so $\overleftrightarrow{XA} \perp m$.

Definition. A point A is a *pole* of a line m if there is a point X on m so that $\overleftrightarrow{AX} \perp m$ and $AX = \omega/2$. If A is a pole of m, we call m a *polar* of A.

Note that A can be a pole of m only when $\omega < \infty$ and that when A is a pole of m so is A^*. The terminology is suggested by the sphere, where if A is the north (or south) pole and m is the equator, then $\overleftrightarrow{AX} \perp m$ and $AX = \omega/2$ for all $X \in m$.

Theorem 14.5. If there are two different lines through a point A and perpendicular to a line m, then A is a pole of m.

Proof. Suppose that $n_1 \neq n_2$ are lines through A perpendicular to m. Say n_1 meets m at X_1 and n_2 meets m at X_2. If $AX_1 = \omega$ then A is on m by Theorem 10.8, a contradiction. So $AX_1 < \omega$. If $X_1 = X_2$ or X_2^*, then X_1 is on n_2 (by Theorem 10.8 if $X_1 = X_2^*$), and hence n_1, n_2 meet in A and in X_1. Then Axiom I4 yields $AX_1 = \omega$, a contradiction. So $X_1 \neq X_2$ or X_2^*. Now A, X_1, X_2 are noncollinear (Figure 14.7).

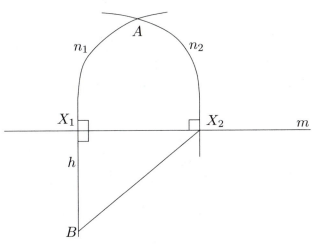

Figure 14.7

Let h be the ray in $\boldsymbol{P}_{X_1}$ that is opposite $\overrightarrow{X_1A}$ (so that $n_1 = h \cup \overrightarrow{X_1A}$). There is a point B on h with $X_1B = X_1A$ by Axiom RR. The Opposite Ray Theorem implies that $B \neq A$. B, X_1, X_2 are noncollinear.

$AX_1 = BX_1, X_1X_2 = X_1X_2$ and $\angle BX_1X_2 = 90 = \angle AX_1X_2$ imply that $\triangle AX_1X_2 \cong \triangle BX_1X_2$, by Axiom SAS. Thus $\angle AX_2X_1 = \angle BX_2X_1$. But $\angle AX_2X_1 = 90$ since $n_2 \perp m$ by hypothesis. So $\overleftrightarrow{BX_2} \perp m$ at point X_2. Then Theorem 14.3 implies that $\overleftrightarrow{BX_2} = n_2$.

So n_1 and n_2 meet in points A and B. Then $AB = \omega$ by Axiom I4. In particular, $\omega < \infty$. Now A-X_1-B (by the Antipode-on-Line Theorem) means that $AX_1 + X_1B = AB = \omega$. But $AX_1 = X_1B$; hence $AX_1 = \omega/2$.

Theorem 14.6. If A is a pole of m, then every line through A is perpendicular to m and meets m at a point distance $\omega/2$ from A. Also, every line perpendicular to m passes through A. So $n \perp m$ if and only if $A \in n$.

Proof. Suppose X is a point of m so that $\overleftrightarrow{AX} \perp m$ and $AX = \omega/2$. Let n be any line through A with $n \neq \overleftrightarrow{AX}$. Then n meets m at some point Y (Theorem 10.11), and A, X, Y are noncollinear (Figure 14.8).

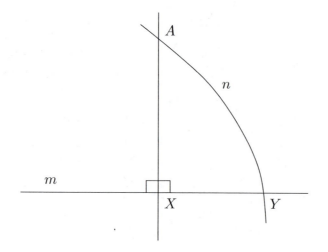

Figure 14.8

Since A-X-A^*, we have $A^*X = \omega/2$ and $\overrightarrow{XA^*} = \overrightarrow{XA}'$. Thus $\angle YXA^* = \angle YXA = 90$. So $\triangle YXA \cong \triangle YXA^*$ by Axiom SAS. Then $AY = A^*Y$ and $\angle AYX = \angle A^*YX$. But $\underline{\angle AYX}$ and $\underline{\angle A^*YX}$ are supplementary, by Corollary 9.8 and Theorem 14.1. Hence, $\angle AYX = 90$ and $n \perp m$. Then $AY = AX = \omega/2$ by *pons asinorum* (Theorem 13.2). This proves the first assertion of the theorem.

If $l \perp m$ at point Z, we have just proved that $\overleftrightarrow{AZ} \perp m$. Hence, $l = \overleftrightarrow{AZ}$ by Theorem 14.3 applied to Z. This establishes the rest of the theorem.

Corollary 14.7. Suppose $\omega < \infty$. Each line m has exactly two poles A and A^*, and each point A has exactly one polar m.

Proof. Problem 9.

The next result will be useful when we study circles in Chapter 21.

Theorem 14.8. Suppose that m is a line and P is a point not on m. Fix a real number r with $0 < r < \omega$. If there are three or more distinct points on m that are distance r from P, then $\omega < \infty, r = \omega/2$, and P is a pole of m.

Proof. We may denote by A, B, C three given distinct points on m with $PA = PB = PC = r$. If A, B, C have no betweenness relation, then by Theorem 8.10 and Theorem 9.7, A and C lie on opposite rays with endpoint B. If A, B, C have a betweenness relation, then by choice of notation we may assume A-B-C. So in any case, we may assume that A and C are on opposite rays with endpoint $B, A \neq B^*$, and $C \neq B^*$ (see Theorems 9.6 and 9.1) (Figure 14.9).

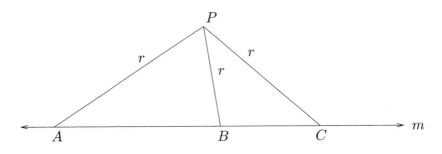

Figure 14.9

Consider the case where $C \neq A^*$. Then we have three isosceles triangles, $\triangle APB, \triangle APC$, and $\triangle BPC$. It is not hard to show that $\angle PBC = \angle PCB = \angle PAB = 90$. The details are left for Problem 10. It follows by Theorem 14.5 that P is a pole of m.

The proof when $C = A^*$ is left for Problem 11.

Definition. The *perpendicular bisector* of a segment $\overline{AB}$ is the line perpendicular to $\overleftrightarrow{AB}$ at the midpoint M of $\overline{AB}$.

Theorem 14.9.   Every point of the perpendicular bisector of a segment is equidistant from the endpoints of the segment.

Proof. Problem 12.

Theorem 14.10.   (Converse of Theorem 14.9) Suppose that segment $\overline{AB}$ is contained in line m and that line $n(\neq m)$ meets m in the midpoint M of $\overline{AB}$. Suppose that there exists a point X on n, but not on m, so that $AX = BX$. Then $n \perp m$ (at M).

Proof. Problem 13.

Definition. A triangle with exactly one right angle is called a *right triangle*; the side opposite the right angle is called the *hypotenuse* and the other two sides are called the *legs* of the triangle. A triangle is called *birectangular* if it has exactly two right angles and *trirectangular* if all three of its angles are right angles.

Examples in $\mathbb{S}$ (Figure 14.10)

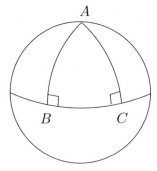

$\triangle ABC$ birectangular

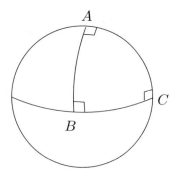

$\triangle ABC$ trirectangular

Figure 14.10

Definition. A triangle is *small* if all of its sides have length less than $\omega/2$ and is *large* if it is not small.

Theorems 14.5 and 14.6 imply that a small triangle can have at most one right angle. If $\omega = \infty$, then every triangle is small, and so there are no birectangular or trirectangular triangles. If $\omega < \infty$, such triangles can always be constructed (see Problem 1).

Problem Set 14

1. Show that if $\omega < \infty$ and $\overline{AB}$ is any segment with $AB \neq \omega/2$, then a birectangular triangle can be found with base $\overline{AB}$ (i.e., with right angles at A and B).

2. Show that if $\triangle ABC$ and $\triangle XYZ$ are trirectangular, then $\triangle ABC \cong \triangle XYZ$.

3. Sketch a right triangle in $\mathbb{S}$ both of whose other two angles are obtuse.

4. Given that $\angle ACM$ is a right angle and that M is the midpoint of both $\overline{AB}$ and $\overline{CD}$, show that $\angle BDM$ is a right angle.

5. Given a proper angle $\underline{hk}$, h' the ray opposite h, that ray r bisects $\underline{hk}$ and ray s bisects $\underline{h'k}$, show that $rs = 90$. (Hint: You will need to show that r-k-s. (Use Insertion.))

6. Show that if h, j, k are rays with h-j-k, and if h', j', k' are their opposite rays, respectively, then h'-j'-k'.

7. Suppose that h, j, k, are rays in $\boldsymbol{P}_Q$ such that $hj = jk = hk = 120$. Let $A \in h$, $B \in j$, $C \in k$ such that $QA = QB = QC = d < \omega$. Finally, assume that A, B, C are all on a line m. Prove that $\omega < \infty$ and $d = \omega/2$. (Hint: Show that no betweenness relation can exist among A, B, and C, and hence that $B \in \overrightarrow{AC}'$.)

8. Show that equilateral triangles exist in any absolute plane. (Hint: Start with any point Q. Use Axiom RF to show that there are rays h, j, k in $\boldsymbol{P}_Q$ so that $hj = jk = hk = 120$. Then

show that there are suitable points A, B, C on h, j, k respectively so that $\triangle ABC$ is equilateral. Problem 7 must play a minor but essential role.)

9. Prove Corollary 14.7.

10. Prove the claim made in the proof of Theorem 14.8 that $\angle PBC = \angle PCB = \angle PAB = 90$.

11. Prove Theorem 14.8 in the case when $C = A^*$.

12. Prove Theorem 14.9.

13. Prove Theorem 14.10.

14. The *rectangle* $ABCD$ with vertices A, B, C, D is $\overline{AB} \cup \overline{BC} \cup \overline{CD} \cup \overline{DA}$, where $\angle ABC = \angle BCD = \angle CDA = \angle DAB = 90$. Prove that there are no rectangles in a plane of finite diameter.

15. Show that if $\omega < \infty$, then each two different lines have a unique common perpendicular. (Hint: If lines m and n have poles M and N, show that $\overleftrightarrow{MN}$ is their unique common perpendicular.)

16. Suppose that A, B, and C are noncollinear points such that $AB = AC = \omega/2$ ($\omega < \infty$). Prove that A is a pole for $\overleftrightarrow{BC}$. (Hint: Consider $\triangle ABC$ and $\triangle A^*BC$.)

17. Assume that $\omega < \infty$, $\overleftrightarrow{XC} \perp \overleftrightarrow{AB}$ at C, A-B-C, $XC = 2\omega/3$, $BC = \omega/4$, D is the midpoint of $\overline{XC}$ (see Figure 14.11).

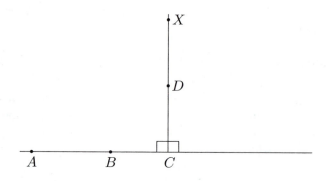

Figure 14.11

For each statement, answer true or false and explain why.

(a) There is a second line through X and perpendicular to $\overleftrightarrow{AB}$.

(b) $\overrightarrow{BA}$-$\overrightarrow{BX}$-$\overrightarrow{BD}$-$\overrightarrow{BC}$.

(c) $\overrightarrow{BX} \cap \overline{AD} = \emptyset$.

(d) There is a line perpendicular to $\overleftrightarrow{AB}$ at B that meets $\overline{XD}$.

(e) There is a point P on $\overrightarrow{CB}'$ so that $\angle DBC = \angle DPC$.

(f) $\triangle DBC \cong \triangle X^*BC$.

15 The Exterior Angle Inequality and the Triangle Inequality

Let $\triangle ABC$ be a triangle in $\mathbb{E}$, and suppose that B-C-D (Figure 15.1).

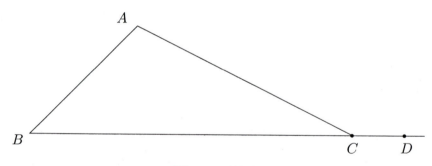

Figure 15.1

A well-known theorem in Euclidean geometry states that

$$\angle ACD = \angle CAB + \angle ABC \qquad (*)$$

and hence that

$$\angle ACD \quad \text{is greater than each of} \quad \angle CAB, \quad \angle ABC. \qquad (**)$$

Now ($**$) is the statement of the Exterior Angle Inequality, which was discussed in Chapter 1. An example was given there that shows that ($*$) and ($**$) are *false* in $\mathbb{S}$. Another example in $\mathbb{S}$ where these statements fail is shown in Figure 15.2.

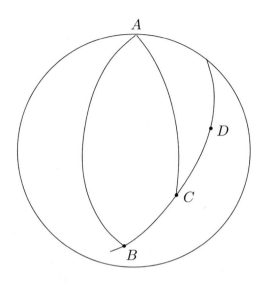

Figure 15.2

Thus neither (*) nor (**) holds in our abstract absolute plane. A version of (**), *under additional hypotheses*, can be proved in the general case and is the content of our Exterior Angle Inequality (Theorem 15.3). First we establish an important preliminary result. Its name stems from that of the Italian geometer Giovanni Ceva (ca. 1647–1736), after whom a segment from a vertex of a triangle to a point on the opposite side is called a *cevian*. Its proof is due to P. Jaskowiak, a student of J. Wetzel.

Theorem 15.1. (Cevian Theorem) Suppose $\omega < \infty$. If $AB < \omega/2$ and $AC \le \omega/2$ in $\triangle ABC$, and if $B\text{-}D\text{-}C$, then $AD < \omega/2$ (Figure 15.3).

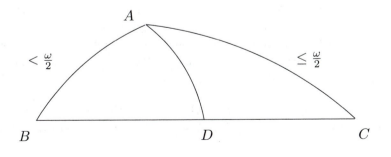

Figure 15.3

Proof. Take $X \in \overrightarrow{AB}$ so that $AX = \omega/2$, and hence $A\text{-}B\text{-}X$ (by Unique Distances for Rays Theorem (8.6)). Let $n \perp \overleftrightarrow{AB}$ at X (Theorem 14.3), and let H be the halfplane with edge n that contains A. Then $B \in H$, by Theorem 10.3 (Figure 15.4).

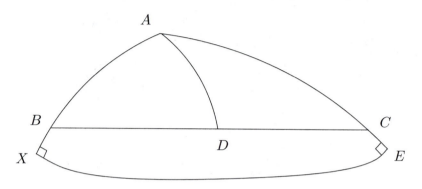

Figure 15.4

Now A is a pole of n, so $\overrightarrow{AC}$ meets n perpendicularly at point E, and $AE = \omega/2$ (Corollary 10.12, Theorem 14.6). So $AC \leq \omega/2$ implies that either $A\text{-}C\text{-}E$ or $C = E$ (Theorem 8.6). If $A\text{-}C\text{-}E$, then $C \in H$ (Theorem 10.3). Then $D \in \overline{BC}$ and convexity of H (Axiom S) yield that $D \in H$. If $C = E$, then $D \in \overline{BE}$ and Theorem 10.3 force $D \in H$. So $D \in H$ in any event.

The ray $\overrightarrow{AD}$ meets n perpendicularly at some point Y with $AY = \omega/2$ (Theorems 10.12 and 14.6). If $A\text{-}Y\text{-}D$, then A and D would be on opposite sides of n, while if $Y = D$, then $D \in n$, a contradiction in either case. Consequently, $A\text{-}D\text{-}Y$, and it follows that $AD < AY = \omega/2$.

The next result contains most of the work for the proof of the Exterior Angle Inequality. If we did not assume $AB \leq \omega/2$ and $BC \leq \omega/2$ (with at least one inequality strict), then we would not necessarily have $BM < \omega/2$ in the following proof. In that case the proof would fail, just as it did in Problem 12 of Chapter 1.

Proposition 15.2. Suppose that, in $\triangle ABC$, $AB \leq \omega/2$ and $BC \leq \omega/2$ with at least one of the inequalities strict. Suppose also that $B\text{-}C\text{-}D$. Then $\angle ACD > \angle CAB$.

Proof. Let M be the midpoint of $\overline{AC}$ (Proposition 8.9). The

Cevian Theorem and $BC, BA \leq \omega/2$ with at least one of these inequalities strict imply that $BM < \omega/2$. Then by Theorem 9.5, there exists a point E with B-M-E and $BM = ME$ (Figure 15.5).

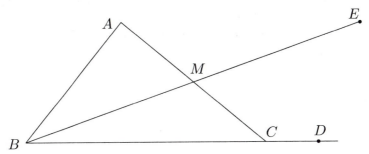

Figure 15.5

The Opposite Ray Theorem implies that $\overrightarrow{MC} = \overrightarrow{MA}'$ (opposite) and $\overrightarrow{ME} = \overrightarrow{MB}'$. So $\angle AMB = \angle CME$ by Theorem 14.2. Now $AM = CM$, $BM = EM$, $\angle AMB = \angle CME$, and Axiom SAS yield $\triangle AMB \cong \triangle CME$. Hence $\angle MCE = \angle MAB = \angle CAB$.

Now B-M-E implies that $\overrightarrow{CB}$-$\overrightarrow{CM}$-$\overrightarrow{CE}$ by Axiom C. $\overrightarrow{CD} = \overrightarrow{CB}'$, by the Opposite Ray Theorem, and hence $\overrightarrow{CB}$-$\overrightarrow{CE}$-$\overrightarrow{CD}$ by Theorem 11.8. So by Theorem 11.5 (Insertion), we have $\overrightarrow{CB}$-$\overrightarrow{CM}$-$\overrightarrow{CE}$-$\overrightarrow{CD}$ and hence $\overrightarrow{CM}$-$\overrightarrow{CE}$-$\overrightarrow{CD}$. Thus

$$\angle ACD = \angle MCD = \angle MCE + \angle ECD \quad > \quad \angle MCE = \angle CAB$$

and the proof is complete.

Definition. Given $\triangle ABC$, let D be a point with B-C-D. Then $\underline{\angle ACD}$ is called an *exterior angle* of $\triangle ABC$, and $\underline{\angle A}$ and $\underline{\angle B}$ are called the *remote interior angles* (relative to $\underline{\angle ACD}$) (Figure 15.6).

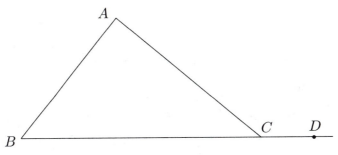

Figure 15.6

Theorem 15.3.   (Exterior Angle Inequality) An exterior angle of a small triangle has larger measure than either remote interior angle.

Proof.   Let $\triangle ABC$ be the small triangle and $\angle ACD$ the given exterior angle, where B-C-D. Then $\angle CAB$ and $\angle ABC$ are the remote interior angles relative to $\angle ACD$.   Proposition 15.2 implies that $\angle ACD > \angle CAB$.

To show that $\angle ACD > \angle ABC$, take any point Q with A-C-Q (Figure 15.7).

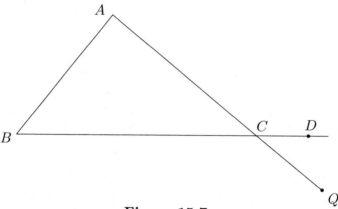

Figure 15.7

Since $\triangle ABC$ is small, Proposition 15.2 also applies to $\angle BCQ$ and yields $\angle BCQ > \angle ABC$. But $\angle BCQ$ and $\angle ACD$ are vertical; hence $\angle BCQ = \angle ACD$ by Theorem 14.2. Therefore, $\angle ACD > \angle ABC$.

Corollary 15.4.   The nonright angles of a small right triangle are acute.

Proof.   Problem 3.

Corollary 15.5.   The base angles of an isosceles triangle whose congruent sides are shorter than $\omega/2$ are acute.

Proof.   Problem 4.

Although the Exterior Angle Inequality in absolute plane geometry requires some special hypotheses, another abstraction of a Euclidean property, the Triangle Inequality, is valid in full generality.

Its proof, however, starts with a special situation.

Proposition 15.6. If $AB < \omega/2$ and $BC \leq \omega/2$ in $\triangle ABC$, then $AB + BC > AC$.

Proof. Suppose (toward a contradiction) that $AB + BC \leq AC$. Then $AC - AB \geq BC$ and $AB < AC$ (as $BC > 0$). So by the Unique Distances for Rays Theorem, there is a point X with A-X-C and $AX = AB$. Then *pons asinorum* applied to $\triangle ABX$ yields $\angle ABX = \angle BXA$ (Figure 15.8).

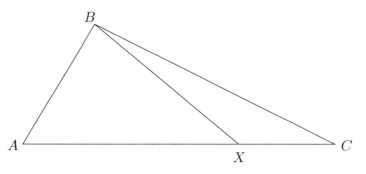

Figure 15.8

$AX + XC = AC$ implies $XC = AC - AX = AC - AB \geq BC$.

Case (1).   Suppose that $XC = BC$. Then *pons asinorum* applied to $\triangle BCX$ implies $\angle XBC = \angle BXC$. Now A-X-C and Axiom C yield $\overrightarrow{BA}$-$\overrightarrow{BX}$-$\overrightarrow{BC}$. Hence

$$\angle ABC = \angle ABX + \angle XBC = \angle BXA + \angle BXC,$$

which equals 180 by the Supplementary Angles Theorem. Then $\overrightarrow{BA}$ and $\overrightarrow{BC}$ are opposite rays, by Axiom M4. So A, B, C are collinear, a contradiction.

Case (2).   Suppose that $XC > BC$. The Unique Distances for Rays Theorem says that there is a point Y with X-Y-C and $CY = BC$. We have $AX = AB$, $CY = CB$, $\angle ABX = \angle AXB$, A-X-C and X-Y-C. Let M be the midpoint of $\overline{XY}$ (see Figure 15.9).

Since $XY < AC$ (Proposition 8.7) and $AC < \omega$,

$$XM = MY = \frac{1}{2}XY < \omega/2.$$

Since $BA < \omega/2$ and $BC \leq \omega/2$ by hypothesis, the Cevian Theorem implies that BX, BM, BY are all less than $\omega/2$. So $\triangle BMX$ and $\triangle BMY$ are small.

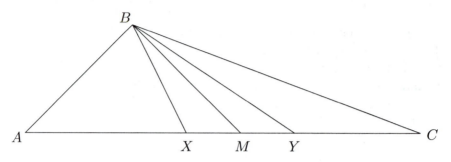

Figure 15.9

Pons asinorum for $\triangle BYC$ yields $\angle CBY = \angle CYB$. Theorem 15.3 applied to $\triangle BMY$ gives $\angle CBY = \angle CYB > \angle CMB$. Applying Theorem 15.3 to $\triangle BMX$ forces $\angle ABX = \angle AXB > \angle BMX$. We add these inequalities and invoke the Opposite Ray Theorem and the Supplementary Angles Theorem to obtain

$$\angle ABX + \angle CBY > \angle BMX + \angle CMB = 180.$$

A-X-C and X-Y-C imply that $\overrightarrow{BA}$-$\overrightarrow{BX}$-$\overrightarrow{BC}$ and $\overrightarrow{BX}$-$\overrightarrow{BY}$-$\overrightarrow{BC}$ (by Axiom C), and hence

$$
\begin{aligned}
\angle ABC &= \angle ABX + \angle XBC \\
&= \angle ABX + \angle XBY + \angle YBC \\
&= (\angle ABX + \angle CBY) + \angle XBY > 180 + \angle XBY > 180.
\end{aligned}
$$

But $\angle ABC > 180$ contradicts Axiom M1.

Since neither Case (1) nor Case (2) can occur, we have reached a contradiction that establishes the result.

Theorem 15.7.   (The Triangle Inequality) In any $\triangle ABC$,

$$AB + BC > AC.$$

Proof.   Since $AC < \omega$, the result holds if $AB \geq \omega/2$ and $BC \geq \omega/2$. So we may assume that one of these, say AB, is less than $\omega/2$. If $BC \leq \omega/2$, then Proposition 15.6 implies that $AB + BC > AC$.

So we may assume that $BC > \omega/2$ and $AB < \omega/2$. In particular, $\omega < \infty$.

If $AC \leq \omega/2$, then $AB + BC > BC > \omega/2 \geq AC$ and we are done. So we may also assume that $AC > \omega/2$. We now have

$$AB < \omega/2, \quad BC > \omega/2, \quad AC > \omega/2.$$

Since $\omega < \infty$, lines $\overleftrightarrow{CB}$ and $\overleftrightarrow{CA}$ meet in C and C^* (Theorem 10.8). Now C-B-C^* by the Antipode-on-Line Theorem, and hence $CB + BC^* = \omega$. Thus $BC^* = \omega - BC$, and similarly $AC^* = \omega - AC < \omega - \omega/2 = \omega/2$.

So in $\triangle ABC^*$, we have $BA < \omega/2$ and $AC^* < \omega/2$. Then Proposition 15.6 implies $BA + AC^* > BC^*$. Hence

$$BA + (\omega - AC) > \omega - BC \quad \Rightarrow \quad AB + BC > AC.$$

The following consequence is also called the Triangle Inequality. Here, points A, B, C may or may not be noncollinear and are not necessarily even distinct.

Corollary 15.8. For any points A, B, C, $AB + BC \geq AC$.

Proof. If A, B, C are noncollinear, this follows from Theorem 15.7; and if A, B, C are distinct and collinear, it is a consequence of Theorem 7.3. If $B = A$ or C, then $AB + BC = AC$; and if $A = C$, then $AB + BC \geq 0 = AC$. So the inequality holds in every case.

Corollary 15.9. If $P_1, P_2, \ldots, P_n$ are any $n \geq 3$ points, then

$$P_1P_2 + P_2P_3 + P_3P_4 + \ldots + P_{n-1}P_n \geq P_1P_n.$$

Proof. By Corollary 15.8, $P_1P_2 + P_2P_3 \geq P_1P_3$. Hence

$$(P_1P_2 + P_2P_3) + P_3P_4 + \ldots + P_{n-1}P_n \geq P_1P_3 + P_3P_4 + \ldots + P_{n-1}P_n,$$

which is $\geq P_1P_4 + P_4P_5 + \ldots + P_{n-1}P_n$ because $P_1P_3 + P_3P_4 \geq P_1P_4$, as before. Continuing (by mathematical induction for complete rigor), we find that $P_1P_4 + P_4P_5 + \ldots + P_{n-1}P_n \geq P_1P_{n-1} + P_{n-1}P_n \geq P_1P_n$, as before.

Problem Set 15

1. Show that if P and Q are any two points on a small triangle $\triangle ABC$, then $PQ < \omega/2$. (Hint: There are three cases: P and Q lie on the same side of the triangle; P is a vertex and Q lies on the opposite side; P and Q are interior points of different sides. Use the Cevian Theorem.)

2. If $AB = AC = \omega/2$ in $\triangle ABC$, prove that A is a pole of $\overleftrightarrow{BC}$.

3. Prove Corollary 15.4.

4. Prove Corollary 15.5. (Hint: Show that if M is the midpoint of the base $\overline{BC}$ of isosceles triangle $\triangle ABC$, then $\triangle ABM$ and $\triangle ACM$ are both small right triangles.)

5. Prove the following converse of Corollary 15.5: If the base angles of an isosceles triangle $\triangle ABC$ are acute, then the equal sides are shorter than $\omega/2$. (Hint: You may assume $\omega < \infty$ (why?). Suppose that $AB = AC \geq \omega/2$ and apply Corollary 15.5 to $\triangle A^*BC$.)

6. Show that for any four points A, B, C, D,

$$AB + BC + CD + DA \geq AC + BD.$$

7. Show that if $\omega < \infty$, then for any triangle $\triangle ABC$,

$$AB + BC + CA < 2\omega.$$

(Hint: Apply the Triangle Inequality to $\triangle BCA^*$.)

8. Show that if $\triangle ABC$ is any small triangle, then $\angle A + \angle B < 180$.

9. Show that if $\triangle ABC$ is any small triangle, then $\angle A + \angle B + \angle C < 270$. (Hint: Problem 8.)

10. (a) Show that if $\triangle ABC$ is any triangle in absolute geometry, then the angle sum $\sigma(ABC) < 540$.

    (b) Find a triangle $\triangle XYZ$ in $\mathbb{S}$ so that $\sigma(XYZ)$ is very close to 540, and a small triangle $\triangle XYZ$ in $\mathbb{S}$ so that $\sigma(XYZ)$ is very close to 270.

11. Let M be the midpoint of $\overline{BC}$ in $\triangle ABC$. Show that

(a) If $AM < w/2$, then $AM < (AB + AC)/2$.

(b) If $AM = w/2$, then $AM = (AB + AC)/2$.

(c) If $AM > w/2$, then $AM > (AB + AC)/2$.

The segment $\overline{AM}$ is called a *median* of $\triangle ABC$. (Hint: For (a), take D on $\overrightarrow{AM}$ so that A-M-D and $AD = 2AM$; for (b), show that $\triangle BAM \cong \triangle CA^*M$; for (c), apply (a) to $\triangle BA^*C$.)

16 Further Results on Triangles

Several comparison theorems for sides and angles of triangles are derived from the Exterior Angle Inequality and the Triangle Inequality. This chapter also includes additional criteria for congruence of triangles, and the basic result on distance between a point and a line (Theorem 16.8).

Theorem 16.1. (Comparison Theorem) If one angle of a triangle is larger than a second, then the side opposite the larger angle is longer than the side opposite the smaller angle; and conversely.

Proof. Assume that $\angle B > \angle C$ in $\triangle ABC$. We want to show that $AC > AB$ (Figure 16.1).

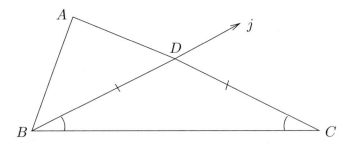

Figure 16.1

By the Unique Angular Distances for Fans Theorem (11.6), there is a ray j with endpoint B so that $\overrightarrow{BC}$-j-$\overrightarrow{BA}$ and (angle measure) $\overline{BC}j = \angle C$. By the Crossbar Theorem (12.4), j meets $\overline{AC}$ at a point

D with A-D-C. Then $\angle DBC \,(= \overrightarrow{BCj}) = \angle ACB = \angle DCB$. So *pons asinorum* (for $\triangle DBC$) implies that $DB = DC$. The Triangle Inequality (Theorem 15.7) applied to $\triangle ABD$ yields $AB < AD + DB = AD + DC = AC$ (the last equality since A-D-C). Thus $AB < AC$.

For the converse, assume that $AC > AB$. We want to show $\angle B > \angle C$. Suppose that $\angle B = \angle C$. Then *pons asinorum* implies that $AC = AB$, a contradiction. Suppose that $\angle B < \angle C$. Then the first part of the proof implies that $AC < AB$, another contradiction. Hence $\angle B > \angle C$.

Corollary 16.2. The hypotenuse of a small right triangle is its longest side.

Proof. Problem 5.

Example. Corollary 16.2 does not hold for all right triangles, as is shown in Figure 16.2 for an example in $\mathbb{S}$.

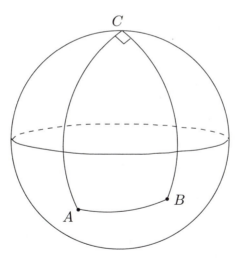

Figure 16.2

The next two results are generalizations of Corollary 16.2. Only one or two of the sides of a triangle with a right angle are assumed to be smaller than $\omega/2$.

Theorem 16.3. Suppose that in $\triangle ABC$, $\angle C = 90$ and $AC < \omega/2$. Then $\underline{\angle B}$ is acute and $AB > AC$.

Proof. If $AB \geq w/2$, then $AB > AC$. So Theorem 16.1 would imply that $90 = \angle C > \angle B$, and we would be done. So we may assume that $AB < w/2$.

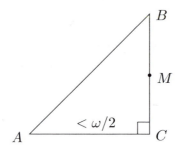

Figure 16.3

Let M be the midpoint of $\overline{BC}$. Then $BM = CM = \frac{1}{2}BC < w/2$. Also, $AM < w/2$ by the Cevian Theorem. So $\triangle ABM$ and $\triangle ACM$ are both small (Figure 16.3).

Corollary 16.2 implies that $AM > AC$. Then Theorem 16.1 (for $\triangle ACM$) yields that $90 = \angle ACM > \angle AMC$. The Exterior Angle Inequality (Theorem 15.3), applied to $\triangle ABM$, gives $\angle B < \angle AMC$. We now have

$$\angle B < \angle AMC < \angle ACM = 90 = \angle C.$$

So $AC < AB$ by Theorem 16.1.

Theorem 16.4. Suppose that in $\triangle ABC, \angle C = 90, AC < w/2$ and $AB < w/2$. Then $\underline{\angle A}$ is acute and $AB > BC$.

Proof. Problem 6.

In general absolute plane geometry, the angle sum of a triangle is not necessarily 180. So knowledge of the measures of two angles of a triangle does not by itself tell you what the third angle measure is. Nevertheless, it turns out that the Angle-Angle-Side criterion for congruence of triangles is valid for small triangles.

Theorem 16.5. (AAS) If in small triangles $\triangle ABC$ and $\triangle XYZ$, $\angle A = \angle X$, $\angle B = \angle Y$ and $BC = YZ$, then $\triangle ABC \cong \triangle XYZ$.

Proof. We will show that $AB = XY$. Once that is accomplished, the conditions $AB = XY$, $BC = YZ$, and $\angle B = \angle Y$ will imply, by Axiom SAS, that $\triangle ABC \cong \triangle XYZ$.

Suppose (toward a contradiction) that $AB > XY$. By the Unique Distances for Rays Theorem, there is a point P so that A-P-B and $BP = XY$ (Figure 16.4).

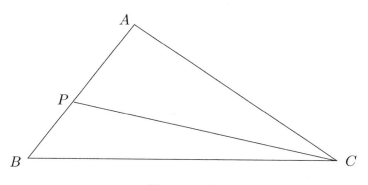

Figure 16.4

Then $BP = YX$, $BC = YZ$, $\angle B = \angle Y$ and Axiom SAS imply that $\triangle BPC \cong \triangle YXZ$. Hence $\angle CPB = \angle ZXY = \angle X = \angle A$.

Now $\triangle ABC$ is small, by hypothesis, so that $CA < \omega/2$ and $CB < \omega/2$. Then $CP < \omega/2$ by the Cevian Theorem (Theorem 15.1). Since $PA < AB < \omega/2$, $\triangle APC$ is small. Thus the Exterior Angle Inequality (Theorem 15.3) implies that $\angle CPB > \angle A$. This contradicts the conclusion that $\angle CPB = \angle A$.

So it is false that $AB > XY$. If $XY > AB$, then the same argument (applied to X, Y, Z in place of A, B, C) gives another contradiction. Therefore, $AB = XY$.

The "hypotenuse-leg" criterion for the congruence of two small right triangles follows from AAS.

Theorem 16.6. Suppose that $\triangle ABC$ and $\triangle XYZ$ are small right triangles with $\angle C = \angle Z = 90$. If $AB = XY$ and $AC = XZ$, then $\triangle ABC \cong \triangle XYZ$.

Proof. Both $BC < \omega/2$ and $YZ < \omega/2$ imply that $BC + YZ < \omega$. By the Unique Distances for Rays Theorem, there is a point P with B-C-P and $BP = BC + YZ$ (Figure 16.5). Then $BP = BC + CP$ implies $CP = YZ$.

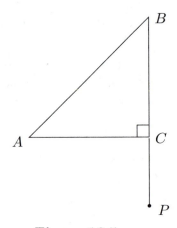

Figure 16.5

Now $\angle ACP = 90$ by the Opposite Ray and Supplementary Angles Theorems. So $AC = XZ$, $CP = ZY$, and $\angle ACP = \angle XZY$ force $\triangle ACP \cong \triangle XZY$ by Axiom SAS. Therefore, $AP = XY = AB$, and $\angle APC = \angle XYZ$.

Then *pons asinorum* (for $\triangle APB$) implies that $\angle ABC = \angle APC = \angle XYZ$. We now have $\angle ABC = \angle XYZ$, $\angle ACB = 90 = \angle XZY$, and $AB = XY$. Thus $\triangle ABC \cong \triangle XYZ$ by Theorem 16.5 (AAS).

Theorem 16.7. (SSA Almost) Suppose that $\triangle ABC$ and $\triangle XYZ$ are triangles with $\angle A = \angle X$, $AB = XY$, and $BC = YZ$. Then either $\triangle ABC \cong \triangle XYZ$ or $\underline{\angle C}$ and $\underline{\angle Z}$ are supplementary.

Proof. Problem 8.

The second alternative of the conclusion of Theorem 16.7 occurs, even in the usual Euclidean plane, as in Figure 16.6.

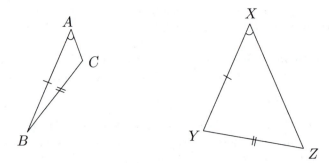

Figure 16.6

Definition.   For any line m and point A, define $d(A, m)$, the *distance between A and m*, to be g.l.b. $\{AX : X \in m\}$, that is, the greatest lower bound of the set of all distances between A and points on m.

Clearly, $d(A, m) = 0$ if A lies on m. If A is not on m, then the story is a little more complicated but is told in full through the next result.

Theorem 16.8.   Suppose that m is a line, C is a point on m, and A is a point not on m so that $\overleftrightarrow{AC} \perp m$.

(a) If $AC < \omega/2$, then $d(A, m) = AC$; and $AC < AX$ for all $X \in m$ with $X \neq C$.

(b) If $AC = \omega/2$ (so in particular, $\omega < \infty$), then $d(A, m) = \omega/2 = AX$ for all $X \in m$.

(c) If $AC > \omega/2$ (so in particular, $\omega < \infty$), then $d(A, m) = \omega - AC = AC^*$, where C^* is the antipode of C.

Proof.   Problem 10.

Suppose that C is a point on line m, A is a point not on m, $\overleftrightarrow{AC} \perp m$, and $AC < \omega/2$. Then C is the closest point on m to A, according to the previous theorem. The next result says that in this situation, as points B on m are taken farther and farther from C (as long as they are on the same ray with endpoint C), then the distances AB increase.

Proposition 16.9.   Suppose in $\triangle ABC$ that $\angle C = 90$, $AC < \omega/2$, and C-D-B. Then $AD < AB$ (Figure 16.7).

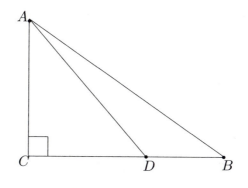

Figure 16.7

Proof. Problem 11.

Theorem 16.10. Suppose in $\triangle ABC$ and $\triangle XYZ$ that $\angle C = 90 = \angle Z$, $AB = XY$, $BC < \omega/2$, $AC < \omega/2$, and $XZ < AC$. Then $YZ > BC$ (Figure 16.8).

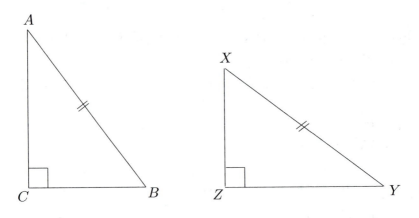

Figure 16.8

Proof. Suppose toward a contradiction that $YZ \leq BC$. Then Theorem 8.6 implies that there is a point D on $\overline{CB}$ with $CD = ZY$ (either C-D-B or $D = B$). Since $XZ < AC$ by hypothesis, there is a point E such that C-E-A and $CE = ZX$ (Figure 16.9).

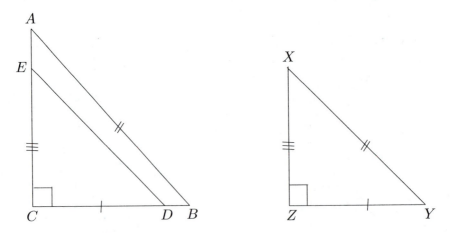

Figure 16.9

Then $\triangle ECD \cong \triangle XZY$ by Axiom SAS. Hence, $ED = XY$.

Since $BC < \omega/2$, we may apply Proposition 16.9 to $\triangle BCA$ to conclude that $BE < BA = XY$. Since $DE = XY$, we must have $D \neq B$, that is C-D-B. Now $EC < AC < \omega/2$, so Proposition 16.9 applies also to $\triangle ECB$. Therefore, $ED < EB < XY$. This contradicts $ED = XY$. Thus, $YZ > BC$.

Problem Set 16

1. Show that if $\overleftrightarrow{AD} \perp \overleftrightarrow{BC}$ at D, $AD < \omega/2$, and $\angle ABC$ and $\angle ACB$ are acute, then B-D-C. (Hints: Consider separately the cases $D = B$, $D = B^*$, D on $\overrightarrow{BC}'$ (toward a contradiction), and D on $\overrightarrow{BC}$ (to obtain B-D-C).)

2. Suppose that $\triangle ABC$ and $\triangle XYZ$ are two small triangles with $\angle A = \angle X$, $AB = XY$, and $\angle B < \angle Y$. Prove that $\angle C > \angle Z$.

3. Suppose that $\triangle ABC$ and $\triangle XYZ$ are two small triangles with $\angle A = \angle X$, $AB = XY$, and $AC < XZ$. Prove that $\angle C > \angle Z$.

4. Suppose that $\triangle ABC$ and $\triangle XYZ$ are two small triangles with $\angle A = \angle X$, $AB < XY$, and $\angle B = \angle Y$. Prove that $AC < XZ$.

5. Prove Corollary 16.2.

6. Prove Theorem 16.4. (Hint: Assume $\angle A \neq \angle X$ and obtain a contradiction.)

7. Find an example of two triangles in $\mathbb{S}$ that satisfy the AAS condition but that are *not* congruent.

8. Prove Theorem 16.7 (SSA Almost). (Hint: Suppose that $XZ > AC$. Apply Theorem 8.6 and show that in this case, $\angle C$ and $\angle Z$ are supplementary.)

9. Show that if $BC < \omega/2$ and $\angle B$ and $\angle C$ are acute, then $\triangle ABC$ is small. (Hint: Let P be the pole of $\overleftrightarrow{BC}$ on the same side of $\overleftrightarrow{BC}$ as A, suppose $\overrightarrow{BA}$ meets $\overrightarrow{PC}$ at X, and apply Theorem 15.1 to $\triangle XBC$.)

10. Prove Theorem 16.8.

11. Prove Proposition 16.9.

12. Prove the **Lemma of Menelaus:**

    Given $\triangle ABC$ and $\omega < \infty$, then

    $$\angle BCA^* > \angle BAC \Leftrightarrow AB + BC < \omega;$$
    $$\angle BCA^* = \angle BAC \Leftrightarrow AB + BC = \omega;$$
    $$\angle BCA^* < \angle BAC \Leftrightarrow AB + BC > \omega.$$

    (Hint: Consider $\triangle A^*BC$ and recall that $\angle BA^*C = \angle BAC$.)

13. Let $\triangle ABC$ be small, and assume that B-D-C and $AC > AB$.

    (a) Order the angles $\angle ABC$, $\angle ADC$, $\angle ACB$ from smallest to largest and prove your answer.

    (b) Prove that $AC > AD$.

    (c) Sketch examples to show that $\angle ABC$ and $\angle ADB$ cannot be compared from the given information.

14. Assume that $\triangle CED$ is small, A-C-D, B-C-E, and $AB > AC$. Prove that $\angle ABC + \angle CED < 180$.

15. Assume that $\angle ABC$ is proper, $AB = CB$, and $\overrightarrow{BA}$-$\overrightarrow{BD}$-$\overrightarrow{BC}$ (Figure 16.10).

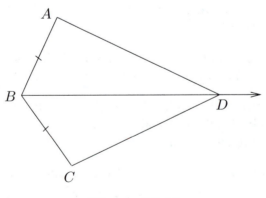

Figure 16.10

    (a) Prove that $\angle ADB$ and $\angle CDB$ are supplementary $\Leftrightarrow D \in \overleftrightarrow{AC}$.

(b) Suppose that $D \notin \overleftrightarrow{AC}$. Prove that $\overrightarrow{BD}$ bisects $\underline{\angle ABC} \Leftrightarrow$ $\angle BAD = \angle BCD$. (Hint: SSA Almost.)

16. Assume the data of Problem 17 in Chapter 14 (Figure 14.11). For each statement, answer true or false and explain why.

   (a) $BD < 7w/12$.

   (b) $\angle XBC < \angle BXC$.

   (c) $\angle DBC < \angle BDX$.

   (d) $BD > w/3$.

   (e) $\underline{\angle XAC}$ is acute.

   (f) $AX > AC$.

17 Parallels and the Diameter of the Plane

We explore in this chapter the relation between parallel lines and the diameter of an absolute plane. Note that our definition of "parallel" applies only to two distinct lines; we may say that a line is parallel to itself, but we do not use this notion here. Furthermore, the definition involves only the intersection of sets. The idea of "parallel" carries no explicit notion of "same slope," which is not defined in our general context.

Definition. Two distinct lines are *parallel* if they are disjoint, that is, if their intersection is empty. If lines m and n are parallel, we write $m \parallel n$.

Suppose a point P is not on a line m. There are exactly three mutually exclusive possibilities:

(I)   There is no line through P parallel to m.

(II) There is exactly one line through P parallel to m.

(III) There are at least two lines through P parallel to m.

Examples. $\mathbb{S}$: (I) holds for every line m and point P not on m.

$\mathbb{E}$: (II) holds for every line m and point P not on m.

$\mathbb{H}$: (III) holds for every line m and point P not on m (Figure 17.1).

Definition. An absolute plane in which (I) holds for *every* line m and point P not on m is called *spherical.*

ℍ:

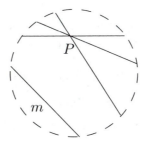

Figure 17.1

Note: An absolute plane that is spherical is sometimes called "doubly elliptic" in the literature. This is because $\omega < \infty$ in such a plane (see the next theorem), and each line has two poles. "Elliptic" is from a Greek work meaning "insufficient"; there are "not enough" parallels.

An absolute plane in which (II) holds for *every* line m and point P not on m is called *Euclidean*.

An absolute plane in which (III) holds for *every* line m and point P not on m is called *hyperbolic* (from a Greek word meaning "excess"; there are "too many" parallels).

It will be proved in Chapter 19 that any absolute plane must be one of spherical, Euclidean, or hyperbolic. In other words, in an absolute plane *it never happens* that one alternative (say, II) occurs for some line m and point P, while another (say, III) occurs for some other line n and point Q. Rather, a plane is uniform: (I) holds throughout, or (II) holds throughout, or (III) holds throughout. But at this point in our development of the theory, we have not yet ruled out the possibility that "mixed behavior" occurs. What we do not prove here (although it is true) is that $\mathbb{S}$, $\mathbb{E}$, and $\mathbb{H}$ are in some sense the only examples of spherical, Euclidean, and hyperbolic planes, respectively. Note that in $\mathbb{G}$, which of course is not an absolute plane, (II) holds for some point and line while (III) holds for some other point and line (see Problem 2).

The relationship between the existence of parallel lines and the diameter ω of the geometry is given in the following theorem. It says that a plane is spherical if and only if it has finite diameter. Notice that the theorem says that three statements are equivalent to one another. If any one of them holds, then so do the other two.

However, none of the statements necessarily has to hold in a given situation, without some further hypothesis.

Theorem 17.1. The following assertions are equivalent.

(a) The diameter ω is finite.

(b) There is a line m and a point P not on m so that no line through P is parallel to m.

(c) For every line m and point P not on m, no line through P is parallel to m.

Proof. We show first that (b) $\Rightarrow$ (a): Assume that m is a line and P a point not on m so that no line through P is parallel to m. (In other words, *all* lines through P *meet* m.) By Theorem 14.4, there is a point Q on m with $\overleftrightarrow{PQ} \perp m$ (at Q). By Theorem 14.3, there is a line n with $n \perp \overleftrightarrow{PQ}$ at P. Now n meets m by assumption, say at point R (Figure 17.2).

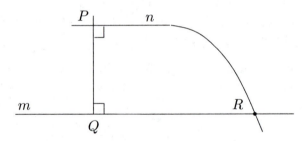

Figure 17.2

Thus $n \neq m$ are both through R and perpendicular to $\overleftrightarrow{PQ}$. Theorem 14.5 implies that $\omega < \infty$.

It follows immediately from Theorem 10.11 that (a) $\Rightarrow$ (c). Since it is trivial that (c) $\Rightarrow$ (b), the proof of Theorem 17.1 is complete.

It follows from Theorem 17.1 that the negations of (a), (b), and (c) are equivalent. We state this formally as follows:

Corollary 17.2. The following assertions are equivalent.

(a)' The diameter $\omega = \infty$.

(b)' For every line m and point P not on m, there is at least one line through P that is parallel to m.

(c)$'$ For some line m and point P not on m, there is at least one line through P that is parallel to m.

The assertion "there is at least one line through P that is parallel to m" is equivalent to "(II) or (III)," where statements (II), (III) are as before.   So Corollary 17.2 shows that

> (II) *or* (III) holds for some line m and point P in $\mathbb{P}$
> $\Leftrightarrow$ (II) *or* (III) holds for all lines m and points P in $\mathbb{P}$
> $\Leftrightarrow$ $\omega = \infty$.

The concept of transversal configuration, which we define next, is important in the analysis of parallel lines in general absolute plane geometry, as well as in the special case of the Euclidean plane.   It also turns out to be useful in situations where lines are not parallel.

Definition. A *transversal configuration* is a triple (t, m, n) of three different lines t, m, n so that t meets m and n in distinct (and not antipodal) points.   The line t is called the *transversal*.

Definition.   Suppose that in a transversal configuration (t, m, n) the transversal t meets m and n in distinct points M and N, respectively.   Let A be a point of m not on t, and let C be a point of n so that A and C lie on the same side of t.   There are points B and D (by Proposition 8.11) so that A-M-B, C-N-D, $AB < \omega$ and $CD < \omega$ (Figure 17.3).

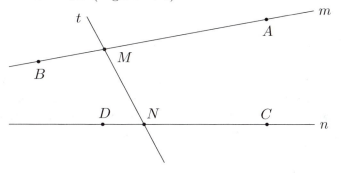

Figure 17.3

Angles $\angle AMN$ and $\angle DNM$, $\angle BMN$ and $\angle CNM$ are called the *alternate interior angles* of (t, m, n). Note that $\angle AMN$ and $\angle BMN$ are supplementary, as are $\angle CNM$ and $\angle DNM$.

Proposition 17.3. Given a transversal configuration (t, m, n) as shown (A and C on the same side of t, etc.) (Figure 17.4),

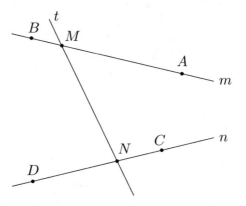

Figure 17.4

then each of the following holds:

(a) $\angle AMN + \angle CNM < 180 \Leftrightarrow \angle BMN + \angle DNM > 180$.

(b) If $\angle AMN + \angle CNM < 180$, then one of $\angle AMN$ or $\angle CNM$ is less than both $\angle BMN$ and $\angle DNM$.

(c) $\angle AMN = \angle DNM \Leftrightarrow \angle BMN = \angle CNM \Leftrightarrow \angle AMN + \angle CNM = 180 \Leftrightarrow \angle BMN + \angle DNM = 180$.

Proof. Problem 1.

Proposition 17.4. Given lines t and m, which intersect at point M, and given N a point on t but not on m, then there exists a unique line n through N such that (t, m, n) is a transversal configuration with a pair of congruent alternate interior angles (Figure 17.5).

Proof. Let A, B be points on m with A-M-B and $AB < \omega$ (Proposition 8.11).

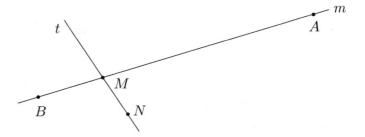

Figure 17.5

By the Opposite Ray and Supplementary Angles Theorems, $\angle AMN$ and $\angle BMN$ are supplementary. In particular, $0 < \angle AMN < 180$. Corollary 12.3 implies that there exist exactly two rays j and k in $\boldsymbol{P}_N$ such that $j\overrightarrow{NM} = k\overrightarrow{NM} = \angle AMN$, and the interior points of j and k lies in opposite halfplanes with edge $\overleftrightarrow{NM}$. So we may assume that $j = \overrightarrow{NU}$, where U and B are on the same side of $\overleftrightarrow{NM}$ (Figure 17.6).

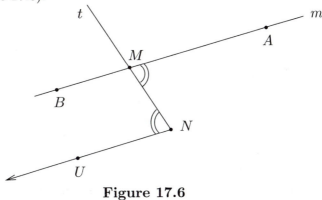

Figure 17.6

Let n be the carrier of j. Then $\angle UNM = j\overrightarrow{NM} = \angle AMN$. So we have a transversal configuration with congruent alternate interior angles. Since j is unique such that $j \in \boldsymbol{P}_N$, $j\overrightarrow{NM} = \angle AMN$, and the interior points of j are in $H(B,t)$, n must also be unique.

The following result shows the relationship between a pair of parallel lines and a transversal configuration with congruent alternate interior angles in all absolute planes of infinite diameter. That is, if $\omega = \infty$ and (t, m, n) is such a configuration, then $m \parallel n$. The converse is not always true. We can find examples in $\mathbb{H}$ where $m \parallel n$ but (t, m, n) does not have congruent alternate interior angles (see

Problem 7). On the other hand, in a Euclidean absolute plane the converse statement is valid (Theorem 17.7).

Theorem 17.5. Assume that $\omega = \infty$. If (t, m, n) is a transversal configuration with a pair of congruent alternate interior angles, then $m \parallel n$.

Proof. Assume $\omega = \infty$, and let (t, m, n) be any transversal configuration such that $\angle AMN = \angle DNM$ (in the notation as in the definition of alternate interior angles). We wish to prove that $m \parallel n$ (Figure 17.7).

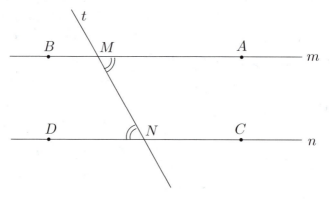

Figure 17.7

Suppose (toward a contradiction) that m and n meet at a point X. A and D are on opposite sides of t, so one of them is on the same side as X. We may assume that A and X are on the same side of t. Then $X \in \overrightarrow{NC}$, and the Opposite Ray Theorem implies that $\overrightarrow{ND} = \overrightarrow{NC}'$. So $\angle DNM$ is an exterior angle for $\triangle MNX$. Since $\omega = \infty$, the Exterior Angle Inequality implies that $\angle DNM > \angle XMN$. But $X \in \overrightarrow{MA}$ yields $\overrightarrow{MA} = \overrightarrow{MX}$ (Theorem 8.4). Hence $\angle XMN = \angle AMN$. So $\angle DNM > \angle AMN$, which contradicts our hypothesis and proves the result.

We can now state Euclid's Fifth Postulate as a property that may or may not hold in a given model of absolute plane geometry. Then Theorem 17.6 will characterize exactly when it does hold.

Euclid's Fifth Postulate. If (t, m, n) is any transversal config- uration so that the measures of two interior angles on the same side (say, on the halfplane H) of t sum to less than 180, then m and n meet in a point of H (Figure 17.8).

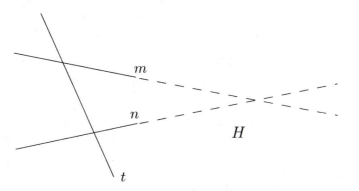

Figure 17.8

When $\omega = \infty$, Axiom I4 implies that two lines meet in *at most one point*. Thus if m and n are as in Euclid's Fifth Postulate and meet in H, then they do *not* meet in the halfplane opposite H.

Theorem 17.6. Assume that $\omega = \infty$. Then an absolute plane $\mathbb{P}$ is Euclidean if and only if Euclid's Fifth Postulate holds for $\mathbb{P}$.

Proof. First, assume that $\mathbb{P}$ is Euclidean. Let (t, m, n) be a transversal configuration with points A-M-B on m, C-N-D on n, and $C \in H = H(A, t)$ (Figure 17.9).

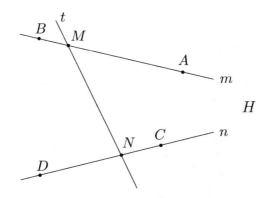

Figure 17.9

We assume that $\angle AMN + \angle CNM < 180$ and will prove that m and n meet in H.

Proposition 17.4 implies that there is a transversal configuration (t, m, l), l through N, which has a pair of congruent alternate interior angles. Since $\omega = \infty$, Theorem 17.5 implies that $l \parallel m$. If $l = n$, then $\angle AMN = \angle DNM$ forces $\angle AMN + \angle CNM = 180$ (Proposition 17.3), which contradicts our hypothesis. So $l \neq n$.

Now l is the unique parallel to m through N, since $\mathbb{P}$ is assumed Euclidean. Thus n is *not* parallel to m. So n meets m, say in a point X.

Suppose (toward a contradiction) that X and B are on the same side of t (opposite H). Then $X \in \overrightarrow{MB} \cap \overrightarrow{ND}$, so $\overrightarrow{MB} = \overrightarrow{MX}$ and $\overrightarrow{ND} = \overrightarrow{NX}$, by Theorem 8.4. So in $\triangle XMN$, $\angle XMN = \angle BMN$ and $\angle XNM = \angle DNM$. Thus

$$\angle AMN > \angle DNM \quad \text{and} \quad \angle CNM > \angle BMN$$

by the Exterior Angle Inequality applied to $\triangle XMN$. This contradicts Proposition 17.3(b).

For the converse, assume that Euclid's Fifth Postulate holds for $\mathbb{P}$. Let m be a line, N any point not on m. We seek to prove that there is a unique line which is through N and parallel to m.

Let M be any point on m, and let $t = \overleftrightarrow{NM}$. Proposition 17.4 implies that there exists a unique line n through N so that the alternate interior angles of the configuration (t, m, n) are congruent. So let A-M-B on m and D-N-C on n, with A and C on the same side of t, as shown in Figure 17.10.

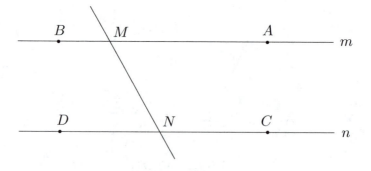

Figure 17.10

So $\angle AMN = \angle DNM$. Theorem 17.5 implies that $n \parallel m$.

Let $l \neq n$ be any other line through N. Let U-N-V on l, with $U \in H(B, t)$. Proposition 17.4 implies that the alternate interior angles of (t, m, l) are *not* congruent. So we may assume that $\angle AMN + \angle VNM < 180$ and $\angle BMN + \angle UNM > 180$ (if not, simply reverse both inequalities). Euclid's Fifth Postulate implies that m meets l. Thus, n is the *unique* parallel to m through N. Hence $\mathbb{P}$ is Euclidean.

Theorem 17.7. Suppose that $\mathbb{P}$ is Euclidean and that (t, m, n) is a transversal configuration such that $m \parallel n$. Then the alternate interior angles of the configuration are congruent.

Proof. Problem 5.

Problem Set 17

1. Prove Proposition 17.3.

2. Find in $\mathbb{G}$ an example of a line m and point P such that (II) holds, and another line m and point P such that (III) holds.

3. Suppose that in a transversal configuration (t, m, n) lines m and n are parallel (so that $\omega = \infty$), and t is perpendicular to m but not to n. Suppose that t meets n at N. Show that there is at least one other line through N that is parallel to m.

4. Show that if $l \parallel m$, then all points of m lie in one of the halfplanes with edge l.

5. Prove Theorem 17.7.

6. Let m be a line and P a point not on m. Suppose that there are lines $t \neq l$, both through P and parallel to m. Let h and k be rays with endpoint P so that $h \subseteq l$, $k \subseteq t$, k^o lies in the halfplane with edge l that does not contain any point of m (see Problem 4), and h^o lies in the halfplane with edge t that contains m (Figure 17.11).

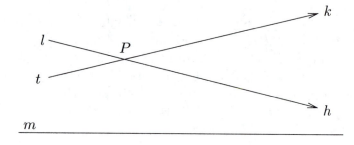

Figure 17.11

If j is any ray with h-j-k, and if n is the carrier of j, show that $n \parallel m$. Hence, prove that there are infinitely many lines through P and parallel to m.

7. Find a transversal configuration (t, m, n) in $\mathbb{H}$ where $m \parallel n$ but the alternate interior angles are not congruent. (Hint: The formula in Chapter 11 for angle measure in $\mathbb{H}$ may be used but it is not necessary.)

8. Assume that $\omega < \infty$. Let (t, m, n) be a transversal configuration where the alternate interior angles are congruent. Let t meet m and n in points M and N, respectively. According to Theorem 10.11, m and n meet in points X and X^*. Prove that neither of $\triangle MNX, \triangle MNX^*$ is small and that MN is the smallest side of both triangles.

9. Given a Euclidean absolute plane with transversal configurations (t, m, n) and (s, m, n) such that $t \parallel s$ and $m \parallel n$. Let t meet m and n in points M and N, respectively, and suppose that s meets m and n in points Q and R, respectively. Prove that $\triangle MQN \cong \triangle RNQ$; hence $MQ = NR$, $MN = QR$, $\angle NMQ = \angle QRN$.

10. Given $m \parallel n$ in a Euclidean absolute plane, let lines t, s meet m at points A, B, respectively, and assume that t, s and n are concurrent at C. Let D, E be on n with D-C-E and A, D on the same side of $\overleftrightarrow{BC}$ (Figure 17.12).

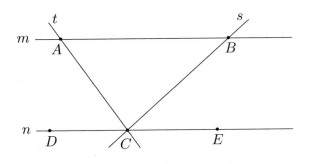

Figure 17.12

(a) Prove that $\overrightarrow{CD}\text{-}\overrightarrow{CA}\text{-}\overrightarrow{CB}\text{-}\overrightarrow{CE}$.

(b) Prove that $\sigma(ABC) = 180$.

18 Angle Sums of Triangles

Recall that for any triangle $\triangle ABC$, the angle sum $\sigma(ABC)$ is defined as $\angle A + \angle B + \angle C$. We have seen examples, in $\mathbb{S}$ for instance, where $\sigma(ABC) \neq 180$. Of course, $\sigma(ABC) < 540$ for all triangles, by Axioms M1 and M4. There are triangles on $\mathbb{S}$ with angle sum arbitrarily close to 540. (See Problem 10 in Chapter 15.) We will show in this chapter that although angle sums are not always constant, their behavior still can be clearly described.

It turns out that when $\omega < \infty$, *all* angle sums of triangles are *greater than* 180 (Theorem 18.1). When $\omega = \infty$, Theorems 18.4 and 18.7 will show that one of two specific situations must occur: *Either all* angle sums of triangles in an absolute plane *equal* 180, *or all* angle sums of triangles in an absolute plane are *less than* 180. The relation between angle sums of triangles and parallel lines will be discussed in the section following this one.

Theorem 18.1. Assume that $\omega < \infty$. Then for all $\triangle ABC$,

$$\sigma(ABC) > 180.$$

Proof. Let $\triangle ABC$ be any triangle. If at least two of the three interior angles have measure 90 or more, then the conclusion follows at once. So it suffices to consider just the case where at least two of the angles are acute. We may assume that $\angle A$ and $\angle B$ are acute. Let X be the pole of $\overleftrightarrow{AB}$ such that X and C are on the same side of $\overleftrightarrow{AB}$ (Corollaries 14.7 and 10.9). Then $XA = XB = \omega/2, \overleftrightarrow{XA} \perp \overleftrightarrow{AB}$ and $\overleftrightarrow{XB} \perp \overleftrightarrow{AB}$ (Theorem 14.6) (Figure 18.1).

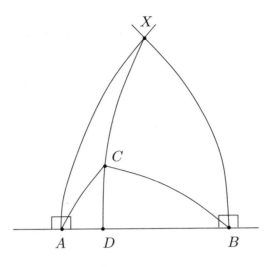

Figure 18.1

Then $\overrightarrow{AX}$ is in the fan $\overrightarrow{ABAC}$ (Theorem 12.2), and $90 = \angle XAB >$ $\angle CAB$ implies that $\overrightarrow{AB}\text{-}\overrightarrow{AC}\text{-}\overrightarrow{AX}$. Thus $\overrightarrow{AB} \in \overrightarrow{AXAC}$, and so $C \in$ $H(B, \overleftrightarrow{AX})$ (the halfplane with edge $\overleftrightarrow{AX}$ that contains B), again by Theorem 12.2. By a similar argument, $C \in H(A, \overleftrightarrow{BX})$. Hence, $C \in$ Int $\angle AXB$. Theorem 12.7 yields that $\overrightarrow{XA}\text{-}\overrightarrow{XC}\text{-}\overrightarrow{XB}$. Then by the Crossbar Theorem (Theorem 12.4), $\overrightarrow{XC}$ meets $\overline{AB}^o$ at some point D. Since X and C are on the same side of $\overleftrightarrow{AB}$, we must have X-C-D (Figure 18.1).

Now $XC < XD = \omega/2, \angle XDB = 90$ (14.6), and $XB = \omega/2$. By Proposition 15.2 applied to $\triangle CBX, \angle DCB > \angle CBX$. Similarly, $\angle DCA > \angle CAX$. Since A-D-B and X-C-D, Axiom C implies that

$$
\begin{aligned}
\angle ACB &= \angle DCB + \angle DCA > \angle CBX + \angle CAX \\
&= (\angle XBD - \angle DBC) + (\angle XAD - \angle DAC) \\
&= (90 - \angle ABC) + (90 - \angle BAC) \\
&= 180 - \angle ABC - \angle BAC.
\end{aligned}
$$

Therefore, $\angle ACB + \angle ABC + \angle BAC > 180$.

Next we explore what happens when $\omega = \infty$. The following result will be applied several times. An informal description of its proof is that a triangle is "placed alongside" a congruent copy of itself to form a "parallelogram."

Proposition 18.2. (Tile-and-Slice) Assume $\omega = \infty$. Given $\triangle ABC$, let l be a line through A such that $(\overleftrightarrow{AC}, l, \overleftrightarrow{BC})$ is a transversal configuration with congruent alternate interior angles. Let D be a point on l such that C and D are on the same side of $\overleftrightarrow{AB}$ and $AD = BC$ (Figure 18.2).

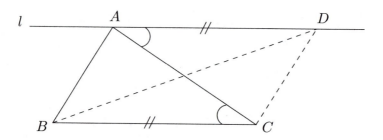

Figure 18.2

Then $l = \overleftrightarrow{AD} \parallel \overleftrightarrow{BC}, \overleftrightarrow{AB} \parallel \overleftrightarrow{CD}, \overrightarrow{AB}\text{-}\overrightarrow{AC}\text{-}\overrightarrow{AD}, \overrightarrow{BA}\text{-}\overrightarrow{BD}\text{-}\overrightarrow{BC}, \overrightarrow{CB}\text{-}\overrightarrow{CA}\text{-}\overrightarrow{CD}, \overrightarrow{DA}\text{-}\overrightarrow{DB}\text{-}\overrightarrow{DC}, \triangle DAC \cong \triangle BCA, \triangle BAD \cong \triangle DCB, \sigma(BAD) = \sigma(ABC)$, and $\angle ABD + \angle ADB = \angle ABC$.

Before we begin the proof, a few remarks are in order. First, the proposition says, among other things, that $\triangle ABD$ has the same angle sum as $\triangle ABC$ but is "flatter": The two angles $\underline{\angle ABD}$ and $\underline{\angle ADB}$ have measures that sum to that of only one angle, $\underline{\angle ABC}$ of $\triangle ABC$. Second, the line l, as given, exists by Proposition 17.4. Third, the point D, as given, exists by Corollary 9.7 (l is the union of opposite rays with endpoint A), Theorem 10.3 (the interior points of opposite rays lie in opposite half-planes with edge $\overleftrightarrow{AB}$), and the Real Ray Axiom. Finally, l is parallel to $\overleftrightarrow{BC}$ (by Theorem 17.5), and it follows from this that A, D, C are noncollinear, as are A, B, D and B, C, D.

Proof. The alternate interior angles $\underline{\angle DAC}$ and $\underline{\angle BCA}$ are congruent, by hypothesis. Since $AD = CB$, by hypothesis, and $AC = CA$, it follows from Axiom SAS that $\triangle DAC \cong \triangle BCA$. Hence $\angle BAC = \angle DCA$ and $AB = CD$.

The argument we now give for line l and $\triangle ABC$ will be used several times in this proof. Note $l \parallel \overleftrightarrow{BC}$ by Theorem 17.5. Hence $\overline{BC} \cap l = \emptyset$, and so B and C are in the same halfplane with edge l (Axiom S). Then, by Theorem 12.2, $\overrightarrow{AC}$ is in the fan $\overrightarrow{AD}\overrightarrow{AB}$. $\overrightarrow{AC} \neq \overrightarrow{AB}$, since A, B, C are noncollinear. If $\overrightarrow{AD}$-$\overrightarrow{AB}$-$\overrightarrow{AC}$, then $\overrightarrow{AB} \cap \overline{CD} \neq \emptyset$ by the Crossbar Theorem. This would place C and D on opposite sides of $\overleftrightarrow{AB}$ (Axiom S), which contradicts our hypothesis. Hence $\overrightarrow{AD}$-$\overrightarrow{AC}$-$\overrightarrow{AB}$, and by Theorem 11.1,

$$\overrightarrow{AB}\text{-}\overrightarrow{AC}\text{-}\overrightarrow{AD}. \tag{$*$}$$

The same argument, applied to line $\overleftrightarrow{BC}$ and $\triangle BAD$, yields $\overrightarrow{BA}$-$\overrightarrow{BD}$-$\overrightarrow{BC}$.

Since $\angle BAC = \angle DCA, (\overleftrightarrow{AC},\overleftrightarrow{AB},\overleftrightarrow{CD})$ is a transversal configuration with congruent alternate interior angles. So $\overleftrightarrow{AB} \parallel \overleftrightarrow{CD}$ by Theorem 17.5. In particular, $\overline{AB} \cap \overleftrightarrow{CD} = \emptyset$. So Axiom S implies that A and B lie on the same side of $\overleftrightarrow{CD}$. Now the argument of $(*)$, applied to line $\overleftrightarrow{CB}$ and $\triangle CDA$, gives us that $\overrightarrow{CB}$-$\overrightarrow{CA}$-$\overrightarrow{CD}$. From line $\overleftrightarrow{DA}$ and $\triangle DCB$ we have $\overrightarrow{DA}$-$\overrightarrow{DB}$-$\overrightarrow{DC}$.

Now

$$\begin{aligned}
\angle BAD &= \angle BAC + \angle CAD &&(\text{since } \overrightarrow{AB}\text{-}\overrightarrow{AC}\text{-}\overrightarrow{AD}) \\
&= \angle DCA + \angle ACB &&(\text{since } \triangle DAC \cong \triangle BCA) \\
&= \angle DCB &&(\text{since } \overrightarrow{CB}\text{-}\overrightarrow{CA}\text{-}\overrightarrow{CD}).
\end{aligned}$$

As $AB = CD$ and $AD = CB$, we have $\triangle BAD \cong \triangle DCB$ by Axiom SAS. Hence $\angle ADB = \angle CBD$, and so

$$\angle ABD + \angle ADB = \angle ABD + \angle CBD = \angle ABC,$$

since $\overrightarrow{BA}$-$\overrightarrow{BD}$-$\overrightarrow{BC}$. Finally,

$$\begin{aligned}
\sigma(BAD) &= \angle BAD + (\angle ABD + \angle ADB) \\
&= (\angle BAC + \angle CAD) + \angle ABC \\
&= \angle BAC + \angle ACB + \angle ABC \ (\text{as } \triangle DAC \cong \triangle BCA) \\
&= \sigma(ABC).
\end{aligned}$$

All the claims of Proposition 18.2 have been proved.

Proposition 18.3. Assume $\omega = \infty$. Given $\triangle ABC$, there exists a triangle $\triangle PQR$ such that $\sigma(PQR) = \sigma(ABC)$ and $\angle Q + \angle R \leq \frac{1}{2}\angle B$.

Proof. $\triangle ABD$, as constructed in Proposition 18.2, has the properties that $\sigma(ABD) = \sigma(ABC)$ and $\angle ABD + \angle ADB = \angle ABC$. Let either $V = B$ and $W = D$, or $V = D$ and $W = B$, so that $\angle AVW$ is the smaller of $\angle ABD$ and $\angle ADB$. Then $\angle AVW \leq \frac{1}{2}\angle ABC$. Now Proposition 18.2, applied to $\triangle AVW$, produces a point E so that $\angle AVE + \angle AEV = \angle AVW$ and $\sigma(AVE) = \sigma(AVW) = \sigma(ABC)$. So we are done by choosing $\triangle PQR = \triangle AVE$.

The next result, together with Theorem 18.1, shows that there never can be two triangles, say $\triangle ABC$ and $\triangle DEF$, in the same absolute plane, with $\sigma(ABC) > 180$ and $\sigma(DEF) \leq 180$. Rather, when $\omega < \infty$, *all* $\sigma(ABC)$ are greater than 180; and when $\omega = \infty$, *all* $\sigma(ABC)$ are less than or equal to 180.

Theorem 18.4. Assume $\omega = \infty$. Then $\sigma(ABC) \leq 180$ for all $\triangle ABC$.

Proof. Suppose (toward a contradiction) that some $\triangle ABC$ satisfies $\sigma(ABC) > 180$. Then $\sigma(ABC) = 180 + p$ for some positive real number p. By Proposition 18.3, there exists $\triangle P_1 Q_1 R_1$ so that $\sigma(P_1 Q_1 R_1) = \sigma(ABC)$ and $\angle Q_1 < \angle Q_1 + \angle R_1 \leq \frac{1}{2}\angle B$. Then the same proposition, applied to $\triangle P_1 Q_1 R_1$, says that there exists $\triangle P_2 Q_2 R_2$ such that

$$\sigma(P_2 Q_2 R_2) = \sigma(P_1 Q_1 R_1) = \sigma(ABC)$$

and

$$\angle Q_2 < \angle Q_2 + \angle R_2 \leq \frac{1}{2}\angle Q_1 < \frac{1}{4}\angle B.$$

For any positive integer n, we apply Proposition 18.3 n times in succession to obtain a $\triangle P_n Q_n R_n$ so that $\sigma(P_n Q_n R_n) = \sigma(ABC)$ and

$$\angle Q_n < \angle Q_n + \angle R_n \leq \frac{1}{2^n}\angle B.$$

Hence,

$$180 + p = \sigma(P_n Q_n R_n) = \angle P_n + \angle Q_n + \angle R_n$$

$$\leq \angle P_n + \frac{1}{2^n}\angle B < 180 + \frac{1}{2^n}\angle B,$$

so that $0 < p \le \dfrac{1}{2^n}\angle B$ for every $n = 1, 2, 3, \ldots$. This is an obvious contradiction for n sufficiently large.

Proposition 18.5.   (Doubling Sides) Suppose that $\triangle ABC$ has $\sigma(ABC) = 180$. Then there exists $\triangle A_1 B_1 C_1$ with $\sigma(A_1 B_1 C_1) = 180, \angle A_1 = \angle A, \angle B_1 = \angle B, \angle C_1 = \angle C, A_1 B_1 = 2AB, B_1 C_1 = 2BC$ and $A_1 C_1 = 2AC$.

Proof. By Theorem 18.1, $\omega = \infty$. As in Proposition 18.2, there is a line l and point D in $l \cap H(C, \overleftrightarrow{AB})$ with $AD = BC$ and $\angle BCA = \angle CAD$. Then $\triangle DAC \cong \triangle BCA$ by Proposition 18.2 (Figure 18.3).

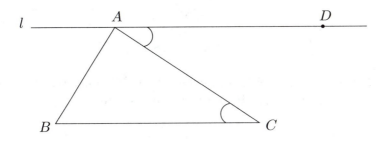

Figure 18.3

By Theorem 9.5 applied to $\overline{BA}$, there exists a point A_1 with B-A-A_1 and $BA_1 = 2AB$. Similarly, there is a point C_1 with B-C-C_1 and $BC_1 = 2BC$. $\overrightarrow{AA_1} = \overrightarrow{AB}'$ by the Opposite Ray Theorem (Figure 18.4).

Now $\overrightarrow{AB}$-$\overrightarrow{AC}$-$\overrightarrow{AD}$ by Proposition 18.2, so Theorem 11.8 and the Rule of Insertion imply $\overrightarrow{AB}$-$\overrightarrow{AC}$-$\overrightarrow{AD}$-$\overrightarrow{AA_1}$. Then D and A_1 are on the same side of $\overleftrightarrow{AC}$ by Theorem 12.1. Also,

$$
\begin{aligned}
180 = \angle BAA_1 &= \angle BAC + \angle CAD + \angle DAA_1 \\
&= \angle BAC + \angle BCA + \angle DAA_1.
\end{aligned}
$$

But by hypothesis $180 = \sigma(ABC) = \angle BAC + \angle BCA + \angle ABC$. Thus $\angle DAA_1 = \angle ABC = \angle CDA$, so that $(\overleftrightarrow{AD}, \overleftrightarrow{AA_1}, \overleftrightarrow{CD})$ is a transversal configuration with congruent alternate interior angles. Since $AA_1 = CD$, Proposition 18.2 applies to line $\overleftrightarrow{AA_1}$ and $\triangle ACD$

to give $\triangle A_1 AD \cong \triangle CDA \cong \triangle ABC$ and $\overrightarrow{DA_1}\text{-}\overrightarrow{DA}\text{-}\overrightarrow{DC}$. In particular, $\angle AA_1 D = \angle BAC$.

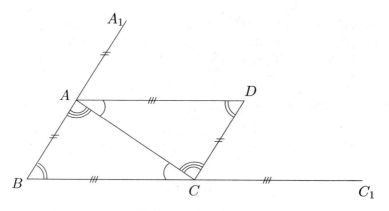

Figure 18.4

A similar argument, applied to line $\overleftrightarrow{CC_1}$ and $\triangle CAD$, implies $\triangle C_1 CD \cong \triangle ADC$ (see Problem 2). In particular, $\angle CC_1 D = \angle DAC = \angle BCA$.

Now we need to show that A_1, D and C_1 are in fact collinear and $A_1\text{-}D\text{-}C_1$. Let $\overrightarrow{DA_1}' = \overrightarrow{DE}$ for a suitable point E. Since $\overrightarrow{DA_1}\text{-}\overrightarrow{DA}\text{-}\overrightarrow{DC}$, then $\overrightarrow{DA_1}\text{-}\overrightarrow{DA}\text{-}\overrightarrow{DC}\text{-}\overrightarrow{DE}$ by the now familiar argument (Theorem 11.8 and the Rule of Insertion). Then

$$
\begin{aligned}
180 = \angle A_1 DE &= \angle A_1 DA + \angle ADC + \angle CDE \\
&= \angle ACB + \angle CBA + \angle CDE
\end{aligned}
$$

(from $\triangle A_1 AD \cong \triangle ABC \cong \triangle CDA$). Hence

$$\angle CDE = 180 - \angle ACB - \angle CBA = \angle BAC,$$

since $\sigma(ABC) = 180$. But $\triangle C_1 CD \cong \triangle ADC$ implies $\angle CDC_1 = \angle DCA = \angle BAC$. Thus $\angle CDE = \angle CDC_1$, so that $\overrightarrow{DE}$ and $\overrightarrow{DC_1}$ form angles of the same measure with $\overrightarrow{DC}$. Now E and A_1 are on opposite sides of $\overleftrightarrow{CD}$ by Theorem 10.3. Since $\overleftrightarrow{AB} \parallel \overleftrightarrow{CD}$, Axiom S implies that A, B, and A_1 are all on the same side of $\overleftrightarrow{CD}$. On the other hand, $B\text{-}C\text{-}C_1$ forces B and C_1 on opposite sides of $\overleftrightarrow{CD}$. Thus

C_1 and E are on the same side of $\overleftrightarrow{CD}$. Corollary 12.3 then implies that $\overrightarrow{DC_1} = \overrightarrow{DE}$.

Hence C_1 is on $\overrightarrow{DA_1}'$ and A_1-D-C_1. Then $A_1C_1 = A_1D + DC_1 = 2AC$ (from $\triangle A_1AD \cong \triangle CDA \cong \triangle DCC_1$). Also, $\overrightarrow{C_1D} = \overrightarrow{C_1A_1}$, $\overrightarrow{A_1D} = \overrightarrow{A_1C_1}$, so that $\angle A_1C_1B = \angle DC_1B = \angle DC_1C = \angle CAD = \angle ACB$ and $\angle BA_1C_1 = \angle AA_1D = \angle BAC$. The result is proved, with $\triangle A_1B_1C_1 = \triangle A_1BC_1$.

Proposition 18.6. (Cutting Down) Assume $\omega = \infty$. If $\triangle ABC$ has $\sigma(ABC) = 180$ and if B-D-C, then $\sigma(ABD) = 180$ (Figure 18.5).

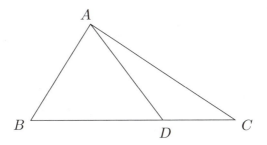

Figure 18.5

Proof. $\overrightarrow{AB}$-$\overrightarrow{AD}$-$\overrightarrow{AC}$ by Axiom C, and hence $\angle BAC = \angle BAD + \angle DAC$. Also, $\overrightarrow{DB} = \overrightarrow{DC}'$ by the Opposite Ray Theorem, and thus $\angle BDA + \angle CDA = 180$ by the Supplementary Angles Theorem. It follows that

$$\sigma(ABD) + \sigma(ACD) = \sigma(ABC) + 180 = 360$$

(see Problem 6). Since $\sigma(ABD)$ and $\sigma(ACD)$ are each at most 180, by Theorem 18.4, the fact that their sum is 360 forces each of them to equal 180. This completes the proof.

Theorem 18.7. Assume that $\mathbb{P}$ is an absolute plane with $\omega = \infty$. If some triangle in $\mathbb{P}$ has angle sum 180, then every triangle in $\mathbb{P}$ has angle sum 180.

Proof. Let $\triangle ABC$ have $\sigma(ABC) = 180$. Let $\triangle DEF$ be any other triangle. If all three of $\angle A, \angle B, \angle C$ are less than or equal to all three of $\angle D, \angle E, \angle F$, then we have $\sigma(ABC) \leq \sigma(DEF)$. Since $\sigma(DEF) \leq$

180 by Theorem 18.4, this forces $\sigma(DEF) = 180$, which is the desired conclusion.

So we may assume that, say, $\angle A > \angle D$. Then the Unique Measures for Fans Theorem (Theorem 11.6) implies that there is a ray j with $\overrightarrow{AB}$-j-$\overrightarrow{AC}$ and $\overrightarrow{AB}j = \angle D$. The Crossbar Theorem (Theorem 12.4) forces j to meet $\overline{BC}$ at a point Q. Thus $\angle BAQ = \angle D$ (Figure 18.6). Also $\sigma(ABQ) = 180$ by Proposition 18.6.

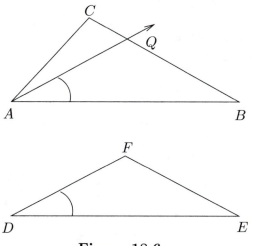

Figure 18.6

By Doubling Sides (Proposition 18.5), there exists $\triangle A_1 B_1 Q_1$ with $\angle A_1 = \angle BAQ = \angle D$, $\angle B_1 = \angle ABQ$, $\angle Q_1 = \angle AQB$, $A_1 Q_1 = 2AQ$, $A_1 B_1 = 2AB$ and $Q_1 B_1 = 2QB$. Then $\sigma(A_1 B_1 Q_1) = \sigma(ABQ) = 180$. We double again, starting with $\triangle A_1 B_1 Q_1$, to obtain a triangle $\triangle A_2 B_2 Q_2$ with $\angle A_2 = \angle A_1$, $\angle B_2 = \angle B_1$, $\angle Q_2 = \angle Q_1$, $A_2 Q_2 = 2A_1 Q_1 = 4AQ$, $A_2 B_2 = 2A_1 B_1 = 4AB$, and $Q_2 B_2 = 2Q_1 B_1 = 4QB$. For any positive integer n, we may repeat the doubling process n times to obtain $\triangle A_n B_n Q_n$ such that

$$\angle A_n = \angle A_1 = \angle D, \qquad \angle B_n = \angle B_1, \qquad \angle Q_n = \angle Q_1$$

so that $\sigma(A_n B_n Q_n) = 180$, and

$$A_n Q_n = 2^n AQ, \qquad A_n B_n = 2^n AB, \qquad Q_n B_n = 2^n QB.$$

There exists an n with $2^n AQ > DF$ and $2^n AB > DE$. Thus $A_n Q_n > DF$ and $A_n B_n > DE$ (Figure 18.7).

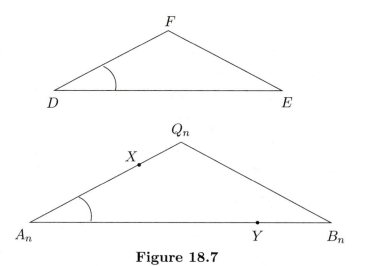

Figure 18.7

Theorem 8.6 (Unique Distances) implies that there is a point X in $\overline{A_n Q_n}$ with $A_n X = DF$, and a point Y in $\overline{A_n B_n}$ with $A_n Y = DE$. Since $\angle X A_n Y = \angle A_n = \angle D$, Axiom SAS forces $\triangle X A_n Y \cong \triangle FDE$.

We apply Cutting Down (Proposition 18.6) to $\triangle Q_n A_n B_n$ to obtain $\sigma(Q_n A_n Y) = 180$. Then by Cutting Down on $\triangle Q_n A_n Y$ we have $\sigma(X A_n Y) = 180$. But $\triangle X A_n Y \cong \triangle FDE$ implies that $\sigma(FDE) = 180$, which is what we wanted to prove.

Problem Set 18

1. Explain how the hypothesis $\omega = \infty$ is used in the proof of Proposition 18.2.

2. Explain how the hypothesis $\sigma(ABC) = 180$ is used in the proof of the Doubling Sides Proposition (Proposition 18.5).

3. Complete the proof of Proposition 18.5 by showing that

$$\triangle C_1 CD \cong \triangle ADC \text{ and } \overrightarrow{DA}\text{-}\overrightarrow{DC}\text{-}\overrightarrow{DC_1}.$$

4. Complete the proof of Proposition 18.6 by showing that

$$\sigma(ABD) + \sigma(ACD) = \sigma(ABC) + 180.$$

5. Given that all three interior angles of $\triangle ABC$ are acute, prove that $\triangle ABC$ must be small. (Hint: We may assume that $\omega < \infty$ and (toward a contradiction) that $BC \geq \omega/2$. Let X be a pole of $\overleftrightarrow{AB}$ so that X and C are on the same side of $\overleftrightarrow{AB}$ (Figure 18.1). If $BC > \omega/2$, let E be the point on $\overline{BC}$ with $BE = \omega/2$. Apply Problem 16 in Chapter 14 or Problem 12 in Chapter 16 to $\triangle XBE$; show that $\triangle XCE$ is small; and apply the Exterior Angle Inequality to $\triangle XCE$ and $\angle XEB$.)

6. Prove the AAS Almost Theorem: Assume that $\omega < \infty$. Let $\triangle ABC$ and $\triangle XYZ$ be triangles such that $\angle A = \angle X, \angle B = \angle Y$ and $BC = YZ$. Then either $\triangle ABC \cong \triangle XYZ$ or $AC + XZ = \omega$. (Hint: It suffices to assume that $\angle C > \angle Z$. (Why?) Then there exists a point D on $\overline{AB}$ with $\triangle DBC \cong \triangle XYZ$. (Why?) Apply Problem 12 in Chapter 16 to $\triangle ADC$.)

7. Assume that $\omega < \infty$. Let A, B, C, P be points such that A-B-P, $AP < \omega/2, \overleftrightarrow{CP} \perp \overleftrightarrow{AB}$ and $CP < \omega/2$. Prove that there exists a point $Q \neq B$ on $\overleftrightarrow{AB}$ with $QC = BC$ and $\angle A^*QC = \angle ABC$. Hence, $\triangle ABC$ and $\triangle A^*QC$ are noncongruent (why?) triangles that satisfy the hypotheses of the AAS Almost Theorem (Problem 6) and where C is not a pole for $\overleftrightarrow{AB}$.

8. Assume that $\omega < \infty$. Suppose that $\triangle ABC$, $\triangle X_1Y_1Z_1$, and $\triangle X_2Y_2Z_2$ are three triangles so that each two of them satisfy the hypotheses of the AAS Almost Theorem (Problem 6). Specifically, assume that $\angle A = \angle X_1 = \angle X_2, \angle B = \angle Y_1 = \angle Y_2, BC = Y_1Z_1 = Y_2Z_2$. Assume also that no two of the triangles are congruent to each other.

   (a) Prove that $AC = X_1Z_1 = X_2Z_2 = \omega/2$. (Hint: Problem 6.)

   (b) Prove that C is a pole for $\overleftrightarrow{AB}$, Z_1 is a pole for $\overleftrightarrow{X_1Y_1}$, and Z_2 is a pole for $\overleftrightarrow{X_2Y_2}$. (Hint: Theorem 16.7.)

   (c) Explain why this problem does not contradict Problem 7.

9. This problem shows that many of the conclusions of Proposition 18.2 remain valid in absolute planes of finite diameter,

even though there are no parallel lines. Assume that $\omega < \infty$. Given $\triangle ABC$, let l be a line through A such that $(\overleftrightarrow{AC}, l, \overleftrightarrow{BC})$ is a transversal configuration with congruent alternate interior angles, and B and C are on the same side of l. Let D be a point on l such that $AD = BC$, and C and D are on the same side of $\overleftrightarrow{AB}$ (Figure 18.8).

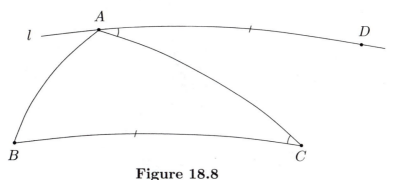

Figure 18.8

(a) Prove that $\triangle DAC \cong \triangle BCA$, hence $BA = CD$.

(b) Prove that $\overrightarrow{AB}$-$\overrightarrow{AC}$-$\overrightarrow{AD}$. (Hint: Apply Theorem 12.2 to two fans.)

(c) Prove that $(\overleftrightarrow{AC}, \overleftrightarrow{AB}, \overleftrightarrow{CD})$ is a transversal configuration with congruent alternate interior angles. (Part of this is to show that B and D are on opposite sides of $\overleftrightarrow{AC}$.)

(d) Prove that A and D are on the same side of $\overleftrightarrow{BC}$. (Hint: Suppose not (toward a contradiction). Show that $\overrightarrow{BC}$ meets $\overline{AD}$ at some point X, and then B-C-X. Use Axiom C to compare $\angle BAX$ with $\angle BCX$ and reach a contradiction.)

(e) Similarly, prove that A and B are on the same side of $\overleftrightarrow{CD}$.

(f) Prove that $\overrightarrow{CB}$-$\overrightarrow{CA}$-$\overrightarrow{CD}$, $\overrightarrow{BA}$-$\overrightarrow{BD}$-$\overrightarrow{BC}$, and $\overrightarrow{DA}$-$\overrightarrow{DB}$-$\overrightarrow{DC}$.

(g) Prove that $\triangle BAD \cong \triangle DCB$.

(h) Prove that $(\overleftrightarrow{BD}, l, \overleftrightarrow{BC})$ is a transversal configuration with congruent alternate interior angles.

(i) Prove that $\sigma(BAD) = \sigma(ABC)$ and $\angle ABD + \angle ADB = \angle ABC$.

(j) Prove that the transversal configuration $(\overleftrightarrow{AB}, l, \overleftrightarrow{BC})$ does *not* have congruent alternate interior angles.

19 Parallels and Angle Sums

We determine in this chapter the relation between parallel lines and angle sums of triangles in absolute plane geometry. It turns out that there are exactly three possibilities; any absolute plane must satisfy one of the following:

(I) the plane is spherical (no parallel lines) *and* all angle sums of triangles are larger than 180; or

(II) the plane is Euclidean (there is a unique parallel to any given line through a given point not on the line) *and* all angle sums of triangles are equal to 180; or

(III) the plane is hyperbolic (there are at least two lines parallel to any given line through a given point not on the line) *and* all angle sums of triangles are less than 180.

We achieve this major result in Theorem 19.4, following several preliminary propositions. The theorem gives us new information even about the example $\mathbb{H}$; namely, that $\sigma(ABC) < 180$ for all triangles $\triangle ABC$ in $\mathbb{H}$.

Note that in Proposition 19.1, the plane is *not* assumed to be necessarily Euclidean. The "unique parallel line" property is not hypothesized to hold throughout the plane, but only with respect to some specific line m and point P not on m.

Proposition 19.1. Assume $\omega = \infty$. If there is a line m, a point P not on m and a *unique* line l through P and parallel to m, then there exists a triangle with angle sum 180.

Proof. Let B and C be two points on m, and let U and R be on l

with U-P-R and R, C on the same side of $\overleftrightarrow{PB}$ (Figure 19.1).

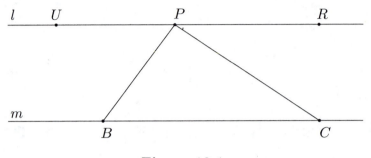

Figure 19.1

Proposition 17.4 implies that there exist transversal configurations $(\overleftrightarrow{PB}, t, m)$ and $(\overleftrightarrow{PC}, u, m)$ with congruent alternate interior angles in each case. Then $t \parallel m$ and $u \parallel m$ by Theorem 17.5. Hence $t = l = u$ by our hypothesis on l. Thus $\angle UPB = \angle CBP$ and $\angle RPC = \angle BCP$.

Theorem 12.2 implies that $\overrightarrow{PR} \in \overrightarrow{PB}\overrightarrow{PC}$. If $\overrightarrow{PB}$-$\overrightarrow{PR}$-$\overrightarrow{PC}$, then $\overrightarrow{PR} \cap \overline{BC} \neq \emptyset$ (Crossbar Theorem), which contradicts $l \parallel m$. Hence $\overrightarrow{PB}$-$\overrightarrow{PC}$-$\overrightarrow{PR}$. The Opposite Ray Theorem, Theorem 11.8, and Insertion yield $\overrightarrow{PU}$-$\overrightarrow{PB}$-$\overrightarrow{PC}$-$\overrightarrow{PR}$. So, by Axiom M4,

$$\begin{aligned} 180 = \angle UPR &= \angle UPB + \angle BPC + \angle RPC \\ &= \angle CBP + \angle BPC + \angle BCP = \sigma(PBC). \end{aligned}$$

Thus $\triangle PBC$ is the desired triangle.

The next proposition says that when $\omega = \infty$, there exists a triangle with a given side, a given sideline, and an arbitrarily small angle opposite the given side.

Proposition 19.2. Assume $\omega = \infty$. Let point P be not on line m and point Q be on line m. Let H be a halfplane with edge $\overleftrightarrow{PQ}$. Then for any positive number ϵ (no matter how small) there exists a point R (depending on ϵ) on m and in H with $\angle PRQ < \epsilon$ (Figure 19.2).

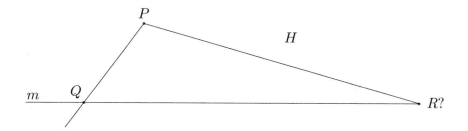

Figure 19.2

Proof. Let E-Q-D on m with $D \in H$. Then $m \cap H$ consists of the interior points of $\overrightarrow{QD}$ (see Theorem 10.3). There exists a point R_1 on $\overrightarrow{QD}$ with $QR_1 = QP$ (by the Real Ray Axiom). Then $\triangle QPR_1$ is isosceles and hence $\angle QPR_1 = \angle QR_1P$ (*pons asinorum*) (Figure 19.3).

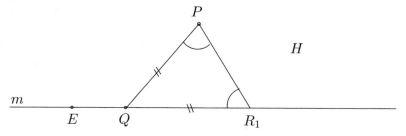

Figure 19.3

Now by Theorem 18.4 and the Supplementary Angles Theorem,

$$180 \geq \sigma(PQR_1) = 2\angle QR_1P + \angle PQR_1$$
$$= 2\angle QR_1P + 180 - \angle EQP.$$

Hence $\angle QR_1P \leq \frac{1}{2}\angle EQP$. Note that $\omega = \infty$ and $R_1 \in \overrightarrow{QD}$ imply E-Q-R_1.

The same argument, starting with points P and R_1 instead of P and Q, produces a point R_2 (Figure 19.4) with Q-R_1-R_2, $R_1R_2 = PR_1$ and

$$\angle QR_2P \leq \frac{1}{2}\angle QR_1P \leq \frac{1}{4}\angle EQP.$$

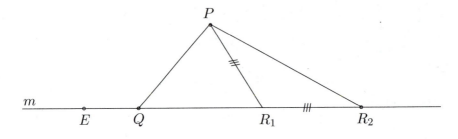

Figure 19.4

Repeating the argument produces a sequence of points in $m \cap H$, say $R_3, R_4, \ldots, R_n, \ldots$ with $\angle QR_nP \le \frac{1}{2^n}\angle EQP$.

Given $\epsilon > 0$, there exists a positive integer n with $\frac{1}{2^n}\angle EQP < \epsilon$. Then the point $R = R_n$ satisfies the required conditions.

The hypothesis of Proposition 19.3 assumes only that two distinct parallel lines exist with respect to some particular line m and point P. It does *not* assume that this "multiple parallel" property must hold throughout the plane. That follows as a consequence of our results.

Proposition 19.3. Assume $\omega = \infty$. If there is a line m, a point P not on m and two distinct lines through P and parallel to m, then there exists a triangle with angle sum less than 180.

Proof. Let Q be any point on m. Proposition 17.4 says that there exists a line l through P such that the transversal configuration $(\overleftrightarrow{PQ}, l, m)$ has congruent alternate interior angles. Then $l \parallel m$ by Theorem 17.5. By hypothesis, there is a line $t \ne l$ through P with $t \parallel m$ (Figure 19.5).

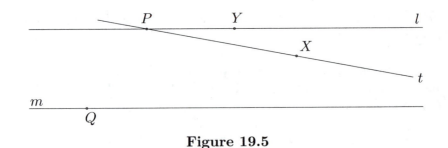

Figure 19.5

Now t consists of a pair of opposite rays with endpoint P, and the interior points of one of these rays lie on the same side of l as does Q. Let X be a point of this ray. Let Y be a point on l and on the same side of $\overleftrightarrow{PQ}$ as is X. Thus $\overrightarrow{PX} \in \overrightarrow{PY}\,\overrightarrow{PQ}$ and $\overrightarrow{PY} \in \overrightarrow{PQ}\,\overrightarrow{PX}$ (Theorem 12.2). It follows that $\overrightarrow{PQ}\text{-}\overrightarrow{PX}\text{-}\overrightarrow{PY}$.

Proposition 19.2 says that there is a point R on m and in $H(X,\overleftrightarrow{PQ})$ with $\angle PRQ < \angle XPY$ (Figure 19.6).

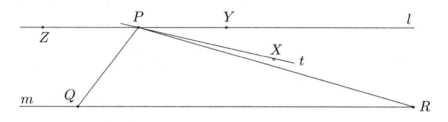

Figure 19.6

Theorem 12.2 implies that $\overrightarrow{PR} \in \overrightarrow{PQ}\ \overrightarrow{PX}$. If $\overrightarrow{PQ}\text{-}\overrightarrow{PX}\text{-}\overrightarrow{PR}$, then $\overrightarrow{PX} \cap \overline{QR} \neq \emptyset$ (Crossbar Theorem), which contradicts $t \parallel m$. Hence $\overrightarrow{PQ}\text{-}\overrightarrow{PR}\text{-}\overrightarrow{PX}$. Then the Rule of Insertion yields $\overrightarrow{PQ}\text{-}\overrightarrow{PR}\text{-}\overrightarrow{PX}\text{-}\overrightarrow{PY}$.

Choose Z on l with Z-P-Y. Our choice of l gives $\angle ZPQ = \angle PQR$. Now $180 = \angle ZPQ + \angle QPY$ (Supplementary Angles Theorem), and $\overrightarrow{PQ}\text{-}\overrightarrow{PR}\text{-}\overrightarrow{PX}\text{-}\overrightarrow{PY}$ implies

$$\angle QPY = \angle QPR + \angle RPX + \angle XPY.$$

Thus

$$\begin{aligned}
180 &= \angle ZPQ + \angle QPY \\
&= \angle PQR + (\angle QPR + \angle RPX + \angle XPY) \\
&> \angle PQR + \angle QPR + \angle XPY \\
&> \angle PQR + \angle QPR + \angle QRP = \sigma(PQR).
\end{aligned}$$

Therefore, $\triangle PQR$ is the desired triangle.

Now we are ready to tackle the main result.

Theorem 19.4. In any absolute plane $\mathbb{P}$, exactly one of the following must occur:

(I) $\mathbb{P}$ is spherical (there are no parallel lines) *and* all angle sums of triangles are larger than 180; or

(II) $\mathbb{P}$ is Euclidean (there is a unique parallel to any given line through a given point not on the line) *and* all angle sums of triangles are equal to 180; or

(III) $\mathbb{P}$ is hyperbolic (there are at least two lines parallel to any given line through a given point not on the line) *and* all angle sums of triangles are less than 180.

Proof. Suppose first that $\omega < \infty$. Then $\mathbb{P}$ is spherical by Theorem 17.1, and all angle sums of triangles exceed 180 by Theorem 18.1. Thus (I) holds in this case.

So we may assume that $\omega = \infty$. If there is some triangle in $\mathbb{P}$ with angle sum 180, then all triangles in $\mathbb{P}$ have angle sum 180 by Theorem 18.7. So $\mathbb{P}$ would have no triangle with angle sum less than 180. Then (the contrapositive of) Proposition 19.3 implies that for any line m and point P not on m, there do not exist two distinct lines through P and parallel to m. There is at least one parallel line to m through P, by Corollary 17.2; hence there is a unique line through P and parallel to m. This argument shows that if there is some triangle with angle sum 180, then (II) holds.

So we may assume that $\omega = \infty$ and there is no triangle in $\mathbb{P}$ with angle sum 180. The contrapositive of Proposition 19.1 implies that for any line m and point P not on m there is not a unique line through P and parallel to m. Corollary 17.2 says that there is at least one line through P and parallel to m, so there must be at least two. All angle sums are less than 180 by Theorem 18.4 and our preceding assumption, so (III) holds.

Problem Set 19

1. Let $\triangle ABC$ be a triangle in an absolute plane $\mathbb{P}$. Let B-C-D, so that $\angle ACD$ is an exterior angle with remote interior angles $\angle ABC, \angle BAC$.

   Prove:

(a)   If $\mathbb{P}$ is spherical, then $\angle ACD < \angle ABC + \angle BAC$.

(b)   If $\mathbb{P}$ is Euclidean, then $\angle ACD = \angle ABC + \angle BAC$.

(c)   If $\mathbb{P}$ is hyperbolic, then $\angle ACD > \angle ABC + \angle BAC$.

2. The number $\delta(ABC) = 180 - \sigma(ABC)$ is called the *defect* of $\triangle ABC$. (So by Theorem 19.4, $\delta(ABC) = 0$ when $\mathbb{P}$ is Euclidean, $\delta(ABC) < 0$ when $\mathbb{P}$ is spherical, and $0 < \delta(ABC) < 180$ for all $\triangle ABC$ when $\mathbb{P}$ is hyperbolic.) Show that the defect is "additive" in the sense that if B-D-C in $\triangle ABC$, then $\delta(ABC) = \delta(ABD) + \delta(ACD)$. (See Proposition 18.6.)

[Because of this, we can use $|\delta(ABC)|$ to *define the area of* $\triangle ABC$ in non-Euclidean planes. When $\omega < \infty$, it is customary to call $\epsilon(ABC) = -\delta(ABC)$ the "excess" of $\triangle ABD$ and to study it instead of the defect.]

3. Assume $\omega = \infty$. Show that if A-X-C and A-Y-B in $\triangle ABC$, if $\overline{BX}$ and $\overline{CY}$ meet at P, and if $\sigma(BPC) = \sigma(ABC)$, then the plane is Euclidean. (Hint: Use the additivity of the defect.)

4. Show that AAA is a criterion for congruence in non-Euclidean planes (i.e., show that if $\angle A = \angle X, \angle B = \angle Y,$ and $\angle C = \angle Z$ in $\triangle ABC$ and $\triangle XYZ$, then $\triangle ABC \cong \triangle XYZ$). (Hence, there is no theory of "similar triangles" in non-Euclidean geometry.) (Hint: If one of the three numbers $AB - XY$, $BC - YZ$, and $CA - ZX$ is zero, the result follows by ASA. Otherwise some two have the same sign, and it entails no loss of generality to suppose that $AB > XY$ and $CA > ZX$. Take points P on $\overline{AB}$ and Q on $\overline{AC}$ so that $AP = XY$ and $AQ = XZ$. Then $\triangle APQ \cong \triangle XYZ$ by SAS. Now use Problem 2 and the fact that $\delta(APQ) = \delta(XYZ) = \delta(ABC)$ to conclude that the plane is Euclidean, contrary to hypothesis.)

5. If M is the midpoint of $\overline{AB}$ in an absolute plane and C is a point not on $\overleftrightarrow{AB}$ with $CM = AM$, prove that

(a) $\angle ACB > 90$ if the plane is spherical.

(b) $\angle ACB = 90$ if the plane is Euclidean (Thales's theorem).

(c) $\angle ACB < 90$ if the plane is hyperbolic.

6. In 1733 Girolamo Saccheri studied the properties of certain quadrilaterals in a little book entitled *Euclides ab omni naevo*

vindicatus, in which he set out to prove the parallel postulate (to "free Euclid of every flaw") but really developed many of the theorems of absolute geometry. In the various parts of this problem we establish some of these properties. Suppose that A and D are on the same side of a line $\overleftrightarrow{BC}$, that $AB = DC < \omega/2$, and that $\angle ABC = \angle DCB = 90$. Then the quadrilateral *ABCD* is called a *Saccheri quadrilateral* with *base* $\overline{BC}$, and angles $\angle BAD$ and $\angle CDA$ are called its *summit angles.*

(a) Show that $AC = BD$.

(b) Prove that $\overrightarrow{BA}$-$\overrightarrow{BD}$-$\overrightarrow{BC}$ and $\overrightarrow{CB}$-$\overrightarrow{CA}$-$\overrightarrow{CD}$. (Hint: To prove that $A \in H(B,\overleftrightarrow{CD})$, argue that otherwise $AB \geq \omega/2$.)

(c) Show that $\triangle ABD \cong \triangle DCA$, and conclude that $\angle BAD = \angle CDA$. So the summit angles of a Saccheri quadrilateral are equal.

(d) Let M be the midpoint of $\overline{AD}$ and N the midpoint of $\overline{CD}$. Show that $\overleftrightarrow{MN}$ is perpendicular to both $\overleftrightarrow{AD}$ and $\overleftrightarrow{BC}$.

(e) Show that $\angle A$ is obtuse, right, or acute accordingly as the plane is spherical, Euclidean, or hyperbolic.

(f) Show that $AD < BC, AD = BC$, or $AD > BC$ accordingly as the plane is spherical, Euclidean, or hyperbolic. (This is a little more difficult than the other parts of this problem.)

7. Let M and N be the midpoints of sides $\overline{AB}$ and $\overline{AC}$ in $\triangle ABC$. Show that $MN > \frac{1}{2}BC, MN = \frac{1}{2}BC,$ or $MN < \frac{1}{2}BC$ accordingly as the plane is spherical, Euclidean, or hyperbolic. (Hint: Let P, Q, R be points on $\overleftrightarrow{MN}$ so that $\overleftrightarrow{AP},\overleftrightarrow{BQ}$, and $\overleftrightarrow{CR}$ are all perpendicular to $\overleftrightarrow{MN}$. Show that $BQRC$ is a Saccheri quadrilateral with base $\overline{QR}$ and that $QR = 2MN$.)

20 Concurrence

The notion of concurrent lines defined in Chapter 5 extends easily to segments and rays. That is, two or more given segments (or rays) are called *concurrent* if there is a point contained in all of them. We observe in this chapter that certain triples of lines or rays associated with triangles often are concurrent. Our context, unless stated otherwise, is an absolute plane $\mathbb{P}$.

Theorem 20.1. The perpendicular bisectors of the three sides of a triangle are concurrent if and only if some two of them meet (Figure 20.1).

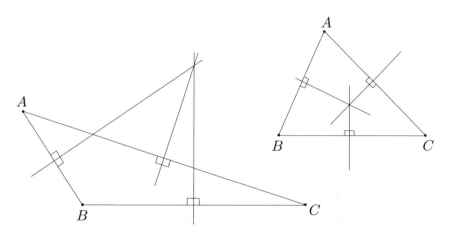

Figure 20.1

Proof. Problem 1.

Since any two lines in a spherical plane meet, it follows from

217

Theorem 20.1 that the perpendicular bisectors of any triangle in a spherical plane are concurrent. It is left as an exercise to show that in a Euclidean plane the perpendicular bisectors of two sides of a given triangle are concurrent, so again the conclusion of Theorem 20.1 holds. But in a hyperbolic plane, there exist triangles in which two perpendicular bisectors are not concurrent.

Example 20.1. Let $O = (0,0)$, $A = (.96,0)$, $B = (0,.96)$ in $\mathbb{H}$. Then $M = (.75,0)$ and $N = (0,.75)$ are the midpoints of $\overline{OA}$ and $\overline{OB}$, resp. (Verification of this is left as an exercise.) Let l be the line through M that is perpendicular to the x-axis in $\mathbb{E}$. So l is the line $x = .75$. Let P, Q be points on l with P-M-Q in $\mathbb{H}$ (Figure 20.2).

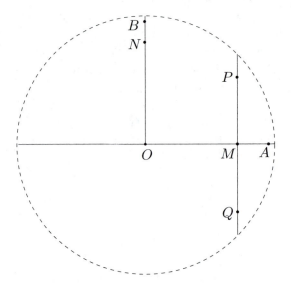

Figure 20.2

Since $\overleftrightarrow{OA}$ (in $\mathbb{H}$) is a diameter of the unit circle, the upper hemisphere placed over the unit circle as in Chapter 11 is cut symmetrically by the upward projection of $\overleftrightarrow{OA}$. Since $\overleftrightarrow{PM} \perp \overleftrightarrow{OA}$ in $\mathbb{E}$, the angles formed by the projections of $\angle PMA$ and $\angle QMA$ onto the hemisphere have equal measure, which must therefore be 90. So by definition of angle measure in $\mathbb{H}$, $\overleftrightarrow{PM} \perp \overleftrightarrow{OA}$ in $\mathbb{H}$. (See Appendix II for a more rigorous discussion of computing angle measure in $\mathbb{H}$.) Thus, the perpendicular bisector of $\overline{OA}$ in $\mathbb{H}$ is the chord of the unit circle contained in the Euclidean line $x = .75$. Similarly, the perpendicular bisector of $\overline{OB}$ in $\mathbb{H}$ is the chord of the unit circle contained in the Euclidean line $y = .75$. It is an easy exercise to check that these two chords do not meet. Hence, the perpendicular bisectors of $\triangle OAB$ in $\mathbb{H}$ are not concurrent.

Definition. The point of concurrence of the three perpendicular bisectors of a triangle (if it exists) is called the *circumcenter* of the triangle. (Of course, in a spherical plane, there will be two such points, a pair of antipodes, for any triangle.)

The connection between the circumcenter of a triangle and circles is made explicit in Chapter 21. The next two results are needed for the study of the bisectors of the interior angles of a triangle.

Theorem 20.2 (Acute Angle Theorem) If $\angle ABC$ is acute, then there is a unique closest point on $\overleftrightarrow{BC}$ to A (call it X); $\overleftrightarrow{AX} \perp \overleftrightarrow{BC}$ and X is on the ray $\overrightarrow{BC}$ (Figure 20.3).

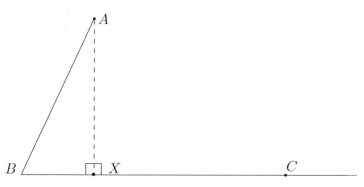

Figure 20.3

Proof. According to Theorem 16.8, the only way there could fail to be a unique closest point on $\overleftrightarrow{BC}$ to A is if $\mathbb{P}$ is spherical, and for all Y on $\overleftrightarrow{BC}$, $AY = \omega/2$ and $\overleftrightarrow{AY} \perp \overleftrightarrow{BC}$. In particular, $\overleftrightarrow{AB} \perp \overleftrightarrow{BC}$, which contradicts $\angle ABC$ acute. So there must be a unique closest point X on $\overleftrightarrow{BC}$ to A. Then $\overleftrightarrow{AX} \perp \overleftrightarrow{BC}$ by Theorem 16.8. Now $X \neq B$ or B^* (if $\omega < \infty$), as $\angle ABC = \angle AB^*C < 90$.

Suppose toward a contradiction that X is on the ray opposite $\overrightarrow{BC}$ (Figure 20.4).

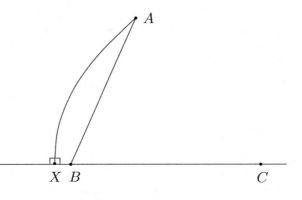

Figure 20.4

Since shortest distance $AX < AB$, the Comparison Theorem (Theorem 16.1) for $\triangle ABX$ yields that $\angle ABX < \angle AXB = 90$. But $\angle ABX$ and $\angle ABC$ are supplementary (Theorem 14.1); hence $\angle ABC > 90$, which contradicts $\angle ABC$ acute. So X cannot be on the ray opposite $\overrightarrow{BC}$ and thus is on $\overrightarrow{BC}$.

Proposition 20.3. If $\triangle ABC$ is a triangle and points U, V are such that B-U-C and A-V-C, then the Cevian segments $\overline{AU}$ and $\overline{BV}$ meet in a point P with B-P-V and A-P-U.

Proof. Problem 5.

Theorem 20.4. The bisectors of the three interior angles of a triangle are concurrent, and their point of intersection is equidistant from all three sides of the triangle.

Proof. Let $\triangle ABC$ be the given triangle. Without loss of generality, we may choose notation so that each side is at least BC in length. By the definition of angle bisector and the Crossbar Theorem (Theorem 12.4), the bisector of $\angle B$ meets $\overline{AC}^o$ at some point Z. Similarly, the bisector of $\angle C$ meets $\overline{AB}^o$ at some point W. By Proposition 20.3, the two bisecting rays meet at a point P with B-P-Z and C-P-W. Thus, we may denote the bisectors of $\angle B$ and $\angle C$ as $\overrightarrow{BP}$ and $\overrightarrow{CP}$, respectively (Figure 20.5).

We complete the proof of the theorem by showing that $\overrightarrow{AP}$ is the bisector of $\angle A$. Note that B-P-Z and Axiom C imply that $\overrightarrow{AB}$-$\overrightarrow{AP}$-$\overrightarrow{AZ}$. Since $\overrightarrow{AZ} = \overrightarrow{AC}$, it follows that $\angle A = \angle BAP + \angle PAC$.

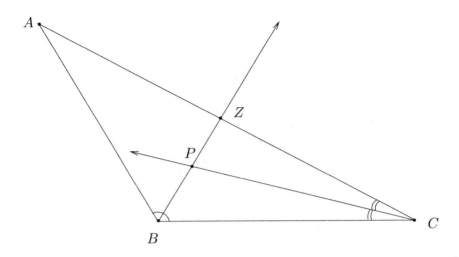

Figure 20.5

Since $\angle PBC = \frac{1}{2}\angle B < 90$, $\underline{\angle PBC}$ (and similarly $\underline{\angle PCB}$) is acute. The Acute Angle Theorem, applied to $\underline{\angle PBC}$, says that the closest point X on $\overleftrightarrow{BC}$ to P lies on $\overrightarrow{BC}$; applied to $\underline{\angle PCB}$, it says that X lies on $\overrightarrow{CB}$ as well. So X is on $\overline{BC}$, with $\overleftrightarrow{PX} \perp \overleftrightarrow{BC}$, and $PX < \omega/2$, by Theorem 16.8 (Figure 20.6).

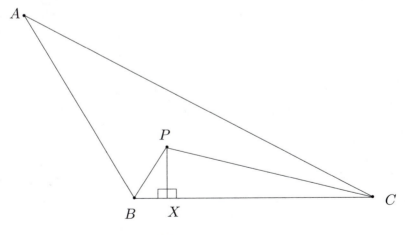

Figure 20.6

Now $BX < BC \leq BA$. So by Theorem 8.6 there is a point U on $\overline{BA}$ with $BU = BX$. Similarly, there is a point V on $\overline{CA}$ with $CV = CX$ (Figure 20.7).

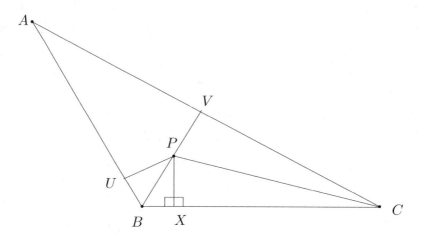

Figure 20.7

Since $\angle PBU = \angle PBA = \angle PBC = \angle PBX$, $\triangle PBU \cong \triangle PBX$ by Axiom SAS. Hence, $\angle PUB = \angle PXB = 90$ and $PU = PX$. Similarly, $\angle PVC = \angle PXC = 90$ and $PV = PX$. Since $PX < \omega/2$, Theorem 16.8 implies that U is the closest point on $\overleftrightarrow{AB}$ to P, and V is the closest point on $\overleftrightarrow{AC}$ to P. Then P is equidistant from $\overleftrightarrow{AB}$, $\overleftrightarrow{BC}$, and $\overleftrightarrow{CA}$, as $PU = PX = PV$.

Since $AP = AP, PU = PV, \angle AUP = 90 = \angle AVP$, the SSA Almost Theorem (Theorem 16.7) implies that either $\triangle AUP \cong \triangle AVP$ or $\underline{\angle UAP}$ and $\underline{\angle VAP}$ are supplementary. But $\underline{\angle UAP} = \underline{\angle BAP}$, $\underline{\angle VAP} = \underline{\angle PAC}$, and as noted previously, $\angle BAP + \angle PAC = \angle A < 180$. So the latter alternative is impossible, and it follows that $\triangle AUP \cong \triangle AVP$. So $\angle UAP = \angle VAP$, hence $\angle BAP = \angle PAC$. Since $\overrightarrow{AB}$-$\overrightarrow{AP}$-$\overrightarrow{AC}$, it follows that $\overrightarrow{AP}$ is the bisector of $\underline{\angle A}$. Therefore, the three angle bisectors meet at P, and the proof is complete.

Definition. The point of concurrence of the three interior angle bisectors of a triangle is called the *incenter* of the triangle. (Since the angle bisectors are rays, a triangle has just one incenter, even when the plane is spherical.)

The connection between the incenter of a triangle and circles is given in Chapter 21.

Several other triples of lines (or rays, or segments) determined by a triangle are concurrent. For example, a *median* is defined as the line through both a vertex of a given triangle and the midpoint of the opposite side. It is a fact that the three medians of any triangle in an absolute plane are concurrent. This is rather difficult to prove in general. A proof for Euclidean planes is given in Chapter 22.

Problem Set 20

1. Prove Theorem 20.1. (Hint: Review the properties of perpendicular bisectors in Chapter 14.)

2. Verify the claims of Example 20.1; that is, show that M, N are the midpoints of $\overline{OA}, \overline{OB}$, respectively, and that the two given chords (the lines $x = .75$ and $y = .75$ restricted to the interior of the unit circle) do not meet.

3. Let $O = (0,0), P_a = (a,0), Q_a = (0,a)$, for each real a with $0 < a < 1$. Determine for exactly which such a the triangle $\triangle OP_aQ_a$ in $\mathbb{H}$ has a circumcenter. (Hint: Compute the midpoint (in $\mathbb{H}$), in terms of a, for $\overline{OP_a}$ and similarly for $\overline{OQ_a}$.)

4. Let $\triangle ABC$ be a triangle in a Euclidean plane, with line l the perpendicular bisector of $\overline{AB}$ and line m the perpendicular bisector of $\overline{BC}$. Prove that l and m meet.

5. Prove Proposition 20.3. (Hint: Crossbar Theorem.)

6. Let $\triangle ABC$ be an isosceles triangle in an absolute plane. Prove that the medians of $\triangle ABC$ are concurrent.

7. Let P be the incenter of $\triangle ABC$ in an absolute plane. Let X be the closest point to P on $\overleftrightarrow{BC}$ (so $X \in \overline{BC}$, as in the proof of Theorem 20.4). Suppose that $\overleftrightarrow{PX}$ is the perpendicular bisector of $\overline{BC}$. Prove that $\triangle ABC$ is isosceles.

8. Let $\triangle ABC$ be equilateral. Prove that it has a circumcenter and that this is the same point as the incenter.

9. Suppose that $\triangle ABC$ is a triangle whose incenter is the same point as its circumcenter. Must $\triangle ABC$ be equilateral? Justify your answer.

21 Circles

The concept of a circle in an absolute plane is defined and studied in this chapter. Many of the familiar properties of Euclidean circles will be seen to remain valid in the more general setting.

Definition. Fix a point P and a number r with $0 < r < w/2$. The *circle* $C = C(P, r)$ with *center* P and *radius* r is defined as the set of all points whose distance from P is r:

$$C(P, r) = \{X : PX = r\}.$$

If $w < \infty$ and we fix a number r with $w/2 < r < w$, then it is an easy exercise to show that

$$\{X : PX = r\} = \{X : P^*X = w - r\} = C(P^*, w - r).$$

Furthermore, it is nearly as easy to see that $\{X : PX = w/2\}$ is a line; namely, the polar of P. So it is no great restriction to constrain the radii of circles to be smaller than $w/2$. Of course, when $w = \infty$ this is no restriction at all.

Note that if A and B are any two distinct points on a circle $C(P, r)$, then $AB \leq AP + PB = 2r < w$. Hence, segment $\overline{AB}$ is defined.

Definition. A *chord* of a circle is a segment $\overline{AB}$, where A and B are any two distinct points on the circle. A *diameter* of a circle is a chord of a circle that contains the circle's center.

Each of the next four propositions is easily proved from the relevant definitions and by the citation of an appropriate result from earlier in the text.

Proposition 21.1.   Let A and B be distinct points on a circle $C(P, r)$. Then $\overline{AB}$ is a diameter if and only if $AB = 2r$.

Proof. Problem 1.

Proposition 21.2.   Let A be any point on a circle. Then there exists one and only one point B on the circle such that $\overline{AB}$ is a diameter.

Proof. Problem 2.

Proposition 21.3.   If $\overline{AB}$ is any chord of a circle $C(P, r)$, then the perpendicular bisector of $\overline{AB}$ contains the center P.

Proof. Problem 3.

Definition. The set of points Int $C(P, r) = \{X : PX < r\}$ is called the *interior* of circle $C(P, r)$. The set $\{X : PX > r\}$ is called the *exterior* of $C(P, r)$.

Proposition 21.4.   If $\overline{AB}$ is any chord of a circle $C(P, r)$, then $PX < r$ for all points X such that A-X-B. Thus, $\overline{AB}^o \subseteq$ Int $C(P, r)$.

Proof. Problem 4.

Theorem 14.8 shows that a line can intersect a circle in at most two distinct points. The next theorem is basic for the further study of how lines and circles meet. It is a result that seems obvious at first glance but is in reality quite deep.

Theorem 21.5.   Let $\overline{AB}$ be a segment and let P be a point with $PA < PB$. Let r be any real number with $PA < r < PB$. Then there exists a point Q on $\overline{AB}$ with $PQ = r$ (Figure 21.1).

The statement of Theorem 21.5, in the case that $r < w/2$, may be rephrased as follows: If one endpoint of a segment lies in the interior of a circle of radius r, and the other endpoint is in the exterior, then the segment meets the circle.

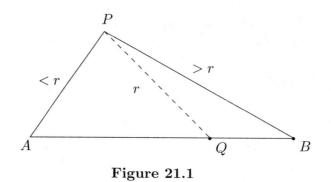

Figure 21.1

One way to prove this result is first to show that as point X runs over $\overline{AB}$, then the distance PX is a continuous function of the distance AX. Since $PA < r$ and $PB > r$, the Intermediate Value Theorem from calculus can be applied, along with the Real Ray Axiom, to prove the existence of the desired point Q. Our proof avoids the explicit use of calculus and proceeds directly from the Least Upper Bound Property. This should be no surprise to experts, since a similar application of the Least Upper Bound Property yields one of the standard proofs of the Intermediate Value Theorem.

Proof of Theorem 21.5. Let T denote the set of points on $\overline{AB}$ that are within distance r of P:

$$T = \{X : X \in \overline{AB} \text{ and } PX \leq r\}.$$

Then we define S as the set of *distances* AX, as X runs over T:

$$S = \{AX : X \in T\}.$$

Since $PA < r$, we have $A \in T$ and $0 = AA \in S$. Certainly, AB is an upper bound for S, since $T \subseteq \overline{AB}$.

The Least Upper Bound Property (see Chapter 4) tells us that S has a least upper bound c. Then $0 \leq c \leq AB$. By Theorem 8.6, there is a point Q on $\overline{AB}$ with $AQ = c$. Let $PQ = q$. We aim to show that $q = r$. Suppose toward a contradiction that $q < r$. Then $PQ < r < PB$ implies that $Q \neq B$.

Hence, either $Q = A$ or A-Q-B, and in either case $c = AQ < AB$. So again by Theorem 8.6, there exists a point X in $\overline{AB}$ with $c < AX < c + (r - q)$. Since $AX > c$, then $AX \notin S$ by definition of least upper bound. Hence $X \notin T$ and so $PX > r$. Now $AQ = c < AX$ implies that A-Q-X or $Q = A$. Therefore,

$$QX = AX - AQ = AX - c < c + (r - q) - c = r - q.$$

Then by the Triangle Inequality,

$$PX \leq PQ + QX = q + QX < q + (r - q) = r.$$

This contradicts $PX > r$ and so proves that $q \not< r$.

Suppose again toward a contradiction that $q > r$. Then $q = PQ > r > PA$ implies that $Q \neq A$ and $Q \notin T$. Also, $c - (q - r) < c$. If Y is in $\overline{AB}$ and $AY > c$, then the definitions of S, and of c as the least upper bound of S, imply that $Y \notin T$. If Y is in $\overline{AB}$ and $c - (q - r) < AY < c$, we must have A-Y-Q, or $Y = A$. Hence,

$$YQ = AQ - AY = c - AY < c - (c - (q - r)) = q - r.$$

Then again by the Triangle Inequality,

$$q = PQ \leq PY + YQ < PY + (q - r).$$

It follows that $PY > r$, and so $Y \notin T$. We have shown that $Y \notin T$ for all Y in $\overline{AB}$ with $AY > c - (q - r)$. This means that $c - (q - r)$ is an upper bound for S, whereas c is the least upper bound. This contradiction implies that $q \not> r$. Since $q \not< r$ and $q \not> r$, we finally have $r = q = PQ$.

Definition. A line that meets a circle in two points is called a *secant* of the circle. A line that meets a circle in precisely one point is called a *tangent* of the circle.

Theorem 21.6. (The Secant Theorem) A line that contains a point A of the interior of a circle meets the circle in exactly two points; that is, it is a secant line of the circle. More precisely, each of the two opposite rays with endpoint A that make up the line meets the circle in some point.

Proof. Suppose that line m contains a point A of the interior of circle $C(P, r)$. So $PA < r$. By Corollary 9.7, m is the union of two opposite rays, say h and k, each with endpoint A (Figure 21.2).

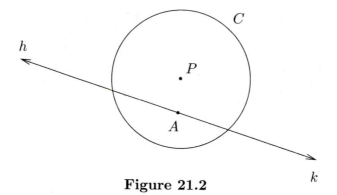

Figure 21.2

Since $2r < \omega$, the Real Ray Axiom tells us that there exists a point B on h with $2r < AB < \omega$. So $B \neq A^*$, and by the Triangle Inequality,

$$2r < AB \leq AP + PB < r + PB.$$

Therefore, $PB > r$. Now by Theorem 21.5, there is a point U in $\overline{AB}$ with $PU = r$. Then $U \in h$ and $U \neq A^*$ (as $AU < AB < \omega$). Similarly, there is a point $V \in k, V \neq A^*$, with $PV = r$. Therefore, U and V are two distinct points in which m meets $C(P,r)$. (Note that m meets the circle only in two points, by Theorem 14.8.) Hence, m is a secant line of $C(P,r)$.

Corollary 21.7. Suppose that line t is tangent to a circle $C(P,r)$ at point A. Then all points other than A on the circle lie in $H(P,t)$, the halfplane with edge t that contains P (Figure 21.3).

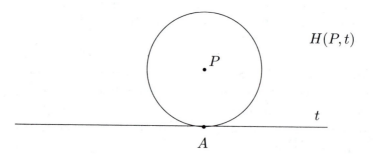

Figure 21.3

Proof. Problem 7.

Theorem 21.8. (The Tangent Theorem) A line t through a point A on $C(P, r)$ is a tangent of the circle if and only if it is perpendicular to $\overleftrightarrow{PA}$ (Figure 21.4).

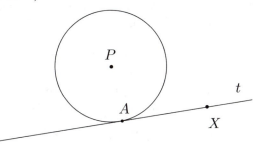

Figure 21.4

Proof. Suppose first that t is a tangent line of $C(P, r)$. Then t contains no interior point of the circle, as otherwise, by the Secant Theorem, t would meet the circle in two points. Hence, every point of t other than A must be an exterior point of the circle. That is, $PX > r = PA$ for all $X \in t$ with $X \neq A$. Since $PA = r < w/2$ by the definition of a circle, it follows from Theorem 16.8 that $\overleftrightarrow{PA} \perp t$.

Now we assume, in order to prove the converse, that $t \perp \overleftrightarrow{PA}$. Suppose toward a contradiction that t is not a tangent of the circle. Then t meets $C(P, r)$ at another point $D \neq A$ (Figure 21.5).

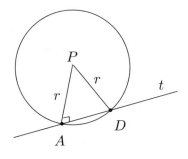

Figure 21.5

Since $PA = r = PD$, $\triangle PAD$ is isosceles. By *pons asinorum*, $\angle PDA = \angle PAD = 90$. Hence, there are two lines through P that are perpendicular to t. By Theorem 14.5, $r = PA = PD = w/2$. But $r < w/2$, by the definition of circle. This contradiction proves that t is a tangent.

It follows immediately from the Tangent Theorem and Theorem 14.3 that for each point A on a circle there is a unique line through A that is tangent to the circle. The next two results deal with the question of whether there exist lines tangent to a circle that pass through a given point in the circle's exterior.

Theorem 21.9. (External Tangent Theorem) Let $C = C(P, r)$ be a circle, and let X be a point in the exterior of C with $PX < \omega - r$. (So if $\omega = \infty$, then PX can be arbitrarily large.) Then there are exactly two lines through X that are tangent to C.

Proof. Let $x = PX$. Then by hypothesis, $x + r < \omega$. Since $r < PX < \omega$, by Theorem 8.6 there exists a point D on $\overline{PX}$ with $PD = r$. That is, $D \in C$. By Theorem 14.3, there is a line l through D that is perpendicular to $\overleftrightarrow{PX}$ (Figure 21.6).

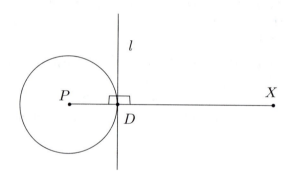

Figure 21.6

Then l is the union of two opposite rays, say h and k, with common endpoint D. By the Real Ray Axiom, there is a point Y on h with $x + r < DY < \omega$. Then by the Triangle Inequality,

$$x + r < DY \leq DP + PY = r + PY.$$

Hence, $PY > x > r = PD$. By Theorem 21.5, applied to segment $\overline{DY}$, point P, and number x, there is a point Z on $\overline{DY}$ such that $PZ = x$. By Theorem 8.6, there is a point B on $\overline{PZ}$ with $PB = r$ (Figure 21.7).

Since $Z, B \notin \overleftrightarrow{PD}$, $\triangle ZPD$, and $\triangle XPB$ are defined. Now $\angle ZPD = \angle XPB$, $PZ = x = PX$, and $PD = r = PB$. Therefore, $\triangle ZPD \cong$

$\triangle XPB$ by Axiom SAS. Hence, $\angle PBX = \angle PDZ = 90$. So $\overleftrightarrow{XB} \perp$ $\overleftrightarrow{PB}$. It follows by the Tangent Theorem that $\overleftrightarrow{XB}$ is tangent to the circle C at B. Note that B and Z are in the same halfplane with edge $\overleftrightarrow{PX}$ (Theorem 10.3).

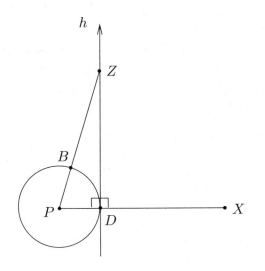

Figure 21.7

A similar argument, applied to the ray k, yields another point Q on C, this time in the opposite halfplane with edge $\overleftrightarrow{PX}$, so that $\overleftrightarrow{XQ}$ is tangent to C at Q (Figure 21.8).

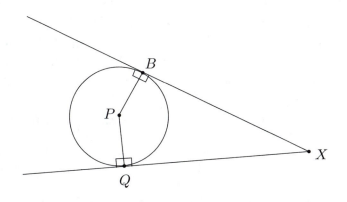

Figure 21.8

We have produced two distinct lines through X that are tangent to C. Thus, points B, Q, X are noncollinear. By Corollary 21.7, all points on C except Q lie in $H(P,\overleftrightarrow{XQ}) = H(B,\overleftrightarrow{XQ})$, and all points on C except B lie in $H(P,\overleftrightarrow{XB}) = H(Q,\overleftrightarrow{XB})$. Therefore, all points on C except B and Q lie in $H(B,\overleftrightarrow{XQ}) \cap H(Q,\overleftrightarrow{XB})$, which is Int $\angle BXQ$ by the definition in Chapter 12.

Suppose toward a contradiction that there is a third line t through X that is tangent to C. Then $t = \overleftrightarrow{XR}$ for some point R on C, $R \neq B$ or Q. Now $R \in$ Int $\angle BXQ$, by the preceding paragraph. So by Theorem 12.7, $\overrightarrow{XB}$-$\overrightarrow{XR}$-$\overrightarrow{XQ}$. Then, according to the Crossbar Theorem (Theorem 12.4), $\overrightarrow{XR}$ contains some point of $\overline{BQ}^o$. This is an interior point of the circle C, by Proposition 21.4. But then the Secant Theorem implies that t is a secant line of C. This contradiction establishes that there are exactly two lines through X that are tangent to C.

Theorem 21.10. Assume that $\omega < \infty$. Let $C = C(P, r)$ be a circle, and let X be a point in the exterior of C.

(a) If $PX = \omega - r$, then there is exactly one line through X that is tangent to C.

(b) If $PX > \omega - r$, then there are no lines through X that are tangent to C.

Proof. (a) Assume $PX = \omega - r$. Then $XP^* = r = X^*P$. That is, X is on $C(P^*, r)$ and X^* is on $C(P, r)$. There is a line l through X^* with $l \perp \overleftrightarrow{PX^*}$ (Theorem 14.3). By the Tangent Theorem, l is the unique line that is tangent to $C(P, r)$ at X^*; and by Theorem 10.8, $X \in l$. Any line m that contains X also passes through X^*. So if such a line m is tangent to $C(P, r)$, then it must be tangent at the point X^*. Therefore, $m = l$. This proves (a).

The proof of (b) is left as an exercise (Problem 10).

The situation of Theorem 21.10(a) is most easily visualized in the spherical model $\mathbb{S}$ (Figure 21.9).

Definition. Let C be a circle, and let $\overline{AB}$ be a diameter of C. Let X be a point on C, with X distinct from A and B. (So A, X, B are noncollinear, by Theorem 14.8.) Then $\angle AXB$ is called an *inscribed angle* (that is, inscribed on the circle as in Figure 21.10).

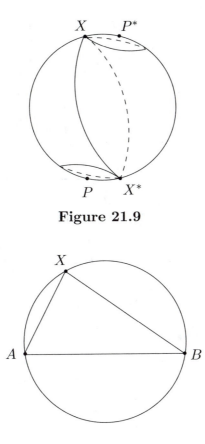

Figure 21.9

Figure 21.10

Theorem 21.11. (Inscribed Angle Theorem) Let $\angle AXB$ be an inscribed angle on a circle. Then $\angle AXB$ is acute if and only if $\mathbb{P}$ is hyperbolic; $\angle AXB$ is obtuse if and only if $\mathbb{P}$ is spherical; and $\angle AXB$ is a right angle if and only if $\mathbb{P}$ is Euclidean.

Proof. Problem 11.

Definition. An *incircle* of a triangle $\triangle ABC$ is a circle for which each of $\overleftrightarrow{AB}, \overleftrightarrow{BC}, \overleftrightarrow{CA}$ is a tangent line, and these lines meet the circle at a point of $\overline{AB}, \overline{BC}$, and $\overline{CA}$, respectively (Figure 21.11).

A *circumcircle* of $\triangle ABC$ is a circle which passes through each of the vertices A, B, C (Figure 21.12).

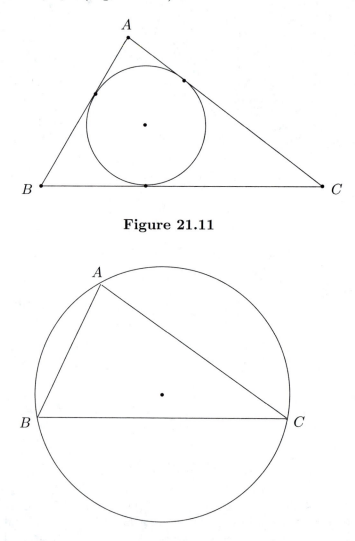

Figure 21.11

Figure 21.12

Theorem 21.12. (a) Every triangle has an incircle.

(b) A triangle has a circumcircle if and only if the perpendicular bisectors of two of the sides meet.

Proof. Problem 12.

The remainder of this chapter deals with the intersection of circles and the application of this to the existence of certain triangles. We begin by studying the following configuration, which is the subject of the next theorem:

Assume that $\overleftrightarrow{PB} \perp \overleftrightarrow{PD}$, with $PB = PD = r < w/2$ (so that B and D are on $C(P,r)$). Let H be the halfplane with edge $\overleftrightarrow{PB}$ that contains D. For each point X on $\overline{PB}$, there is a line l through X with $l \perp \overleftrightarrow{PB}$. Line l is comprised of two opposite rays with endpoint X, and the interior points of one of the rays lie in H. Call this ray h. If $X \neq B$, then $PX < PB = r$. By the Secant Theorem, h meets the circle $C(P,r)$ in a unique point, which we shall call X' (Figure 21.13).

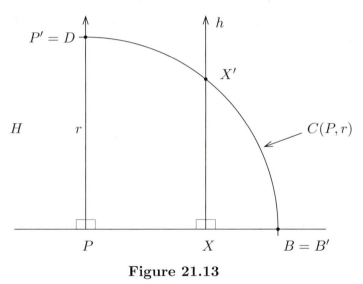

Figure 21.13

This notation will be in effect for the next three results. Before we state the next theorem precisely, we describe informally what it is saying:

Think of the distance XX' as the "height" of the circle directly "above" X. For instance, $PP' = r$, $BB' = BB = 0$. Then two things occur: (a) The farther from P we choose X, then the smaller height XX' becomes; (b) as X runs through all points of $\overline{PB}$, the heights XX' range through all real numbers from r down to 0. Now we explicitly state the theorem.

Theorem 21.13. Assume the hypotheses and notation given previously, including the definition of point X' on $C = C(P, r)$, for each point X in $\overline{PB}$. Then

(a) If $X_1, X_2 \in \overline{PB}$ with $PX_1 < PX_2$ ($\Leftrightarrow X_1 = P \neq X_2$ or $P\text{-}X_1\text{-}X_2$), then $X_1X_1' > X_2X_2'$.

(b) $\{XX' : X \in \overline{PB}\} = [0, r]$.

Proof. Since X' is on $C(P, r)$ for all $X \in \overline{PB}$, $PX' = r$. In particular, $PP' = r$ and $BB' = 0$. For all points X in $\overline{PB}^o, \triangle PXX'$ exists with $\angle X = 90$ and $PX < PB = r = PX' < \omega/2$. By Theorem 16.4, $r = PX' > XX'$. So $0 \leq XX' \leq r$ for all $X \in \overline{PB}$.

(a) Assume $X_1, X_2 \in \overline{PB}$ with $PX_1 < PX_2$. If $X_1 = P$, we have just shown that $X_1X_1' = r > X_2X_2'$. If $X_2 = B$, then $X_1X_1' > 0 = X_2X_2'$. So we need now consider only points $X_1, X_2 \in \overline{PB}^o$. Then we have two triangles, $\triangle PX_1'X_1$ and $\triangle PX_2'X_2$, with $\angle X_1 = \angle X_2 = 90, PX_1' = PX_2' = r, PX_1 < PX_2 < r < \omega/2$, and $X_2X_2' \leq r < \omega/2$.

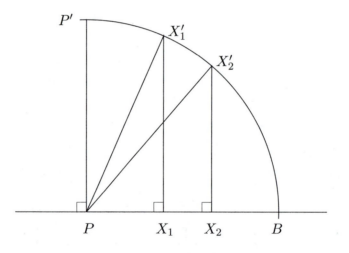

Figure 21.14

Then by Theorem 16.10, $X_1X_1' > X_2X_2'$. The proof of (a) is complete.

(b) Let s be any real number with $0 < s < r$. By Theorem 8.6, there is a point S in $\overline{PP'}^o$ with $PS = s$. Then P is an interior point of the circle $C(S, r)$. By the Secant Theorem, $C(S, r)$ meets $\overrightarrow{PB}$ at some point X; thus $SX = r$ (Figure 21.15).

We need to be sure that X is in $\overline{PB}^o$. But we have $\triangle SXP$ with $\angle P = 90, SP = s < r < \omega/2$ and $SX = r < \omega/2$. Then Theorem 16.4 implies that $SX > XP$. So $PX < r = PB$, which means that $X \in \overline{PB}^o$.

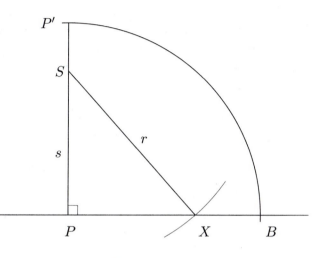

Figure 21.15

Now $\triangle SXP$ and $\triangle X'PX$ are both small, with $\angle SPX = \angle X'XP$ = 90 (Figure 21.16).

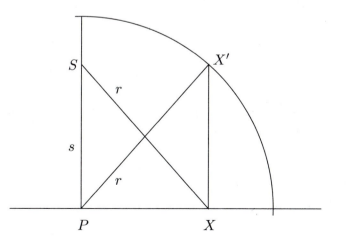

Figure 21.16

Since $SX = r = X'P$ and $XP = PX$, Theorem 16.6 implies that $\triangle SXP \cong \triangle X'PX$. Therefore, $s = SP = X'X$. So the numbers XX', as X runs through $\overline{PB}^o$, cover every number s with $0 < s < r$. In other words, $\{XX' : X \in \overline{PB}\} = [0, r]$ and (b) is proved.

We continue to assume the configuration of Theorem 21.13, with $C = C(P, r)$, and in addition we suppose that there is another circle $C_1 = C(Q, t)$ with center Q on ray $\overrightarrow{PB}$, such that $P\text{-}B\text{-}Q$ and $PQ < \omega$. Furthermore, we assume that $BQ < t < PQ$. By (8.6), there is a point A on $\overrightarrow{QB}$ with $AQ = t$. It follows that A is on C_1 and $P\text{-}A\text{-}B\text{-}Q$ (Figure 21.17).

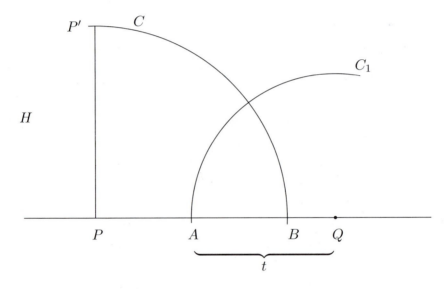

Figure 21.17

For each point X in $\overline{AQ}$, there is a ray h with endpoint X such that h is part of a line l with $l \perp \overleftrightarrow{AQ}$ at X, and h^o lies in the halfplane H. (If $X \in \overline{AB} \subseteq \overline{PB}$, then h is exactly as before.) Then again by the Secant Theorem, h meets C_1 in a unique point, which we shall call X'' (Figure 21.18).

We want to prove that there is a point Y in $\overline{AB}$ so that $Y' = Y''$; in other words, that C and C_1 meet at Y'. This may appear obvious, but its proof takes a lot of work. Like the proof of Theorem 21.5, it is based on the Least Upper Bound Property.

First, we establish a result about the segment $\overline{AB}$. It says roughly that if Y is a point of $\overline{AB}$ such that directly over Y one of the circles is "above" the other, then these relative positions stay the same for at least a little way to either side of Y.

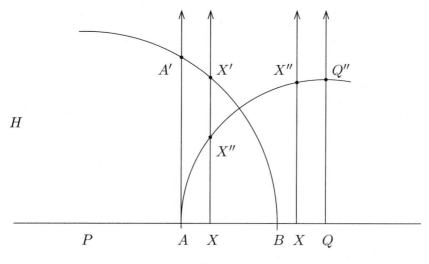

Figure 21.18

Proposition 21.14. Suppose that circles $C = C(P, r)$ and $C_1 = C(Q, t)$, points A, B with P-A-B-Q, and for each $X \in \overline{AB}$, points X' on C and X'' on C_1 are as given.

(a) Suppose that Y is a point in $\overline{AB}$ with $YY'' < YY'$. Then $Y \neq B$, and there is a point Z with Y-Z-B and such that $ZZ'' < ZZ'$.

(b) Suppose that Y is a point in $\overline{AB}$ with $YY'' > YY'$. Then $Y \neq A$ and there is a point W with A-W-Y and such that $WW'' > WW'$.

Proof. We prove (a). The proof of (b) is the same, with the roles of circles C and C_1, and of points A and B, interchanged. So assume that point Y is in $\overline{AB}$, and $YY'' < YY'$. Since $BB = 0$, we know that $Y \neq B$ (Figure 21.19).

By Theorem 21.13, $\{XX' : X \in \overline{AB}\} = [0, AA']$; and if A-X_1-X_2-B, then $X_1X_1' > X_2X_2'$. Since $YY' > YY'' > 0$, there exists a point Z_1 with Y-Z_1-B such that $YY' > Z_1Z_1' > YY''$. (That is, a number between YY' and YY'' must be realized as some height Z_1Z_1'.)

Theorem 21.13, applied to circle C_1 and segment $\overline{AQ}$, tells us that $\{XX'' : X \in \overline{AB}\} = [0, BB'']$; and if A-X_1-X_2-B, then $X_1X_1'' < X_2X_2''$. Hence, there is a point Z with Y-Z-Z_1 such that $YY'' < ZZ'' < Z_1Z_1'$. Now A-Z-Z_1 (by insertion, if $Y \neq A$). So by Theorem 21.13, applied again to circle C, we have $Z_1Z_1' < ZZ'$. Therefore, $ZZ'' < ZZ'$. Lastly, Y-Z_1-B and Y-Z-Z_1 give Y-Z-B, by insertion.

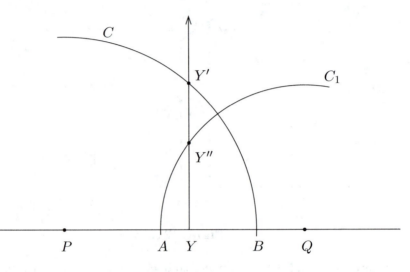

Figure 21.19

Theorem 21.15.   (Circle Intersection Theorem)   Suppose that $C = C(P,r)$ and $C_1 = C(Q,t)$ are two circles such that $max\{r,t\} \leq PQ < r + t$ (where $max\{r,t\}$ means the larger of r and t). Then C and C_1 meet in exactly two distinct points, neither of which is on $\overleftrightarrow{PQ}$.

Proof.   Since $PQ \geq r$, there is a point B in $\overline{PQ}$ with $PB = r$ (Theorem 8.6); and $PQ \geq t$ implies that there is a point A in $\overline{PQ}$ with $QA = t$. Since $PA = PQ - QA < r + t - t = r$, we have P-A-B-Q, unless $P = A$ or $Q = B$ (Figure 21.20).

Choose a halfplane H with edge $\overleftrightarrow{PQ}$. For each point X in $\overline{AB}$, there is a unique ray h with endpoint X such that h^o lies in H and the carrier of h is perpendicular to $\overleftrightarrow{PQ}$ at X. As in the previous discussion, h meets C is a unique point X', and h meets C_1 in a unique point X'' (Figure 21.21).

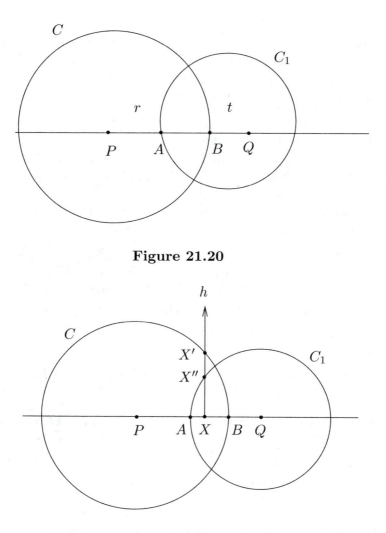

Figure 21.20

Figure 21.21

In particular, $A'' = A \neq A'$ and $B' = B \neq B''$. So $AA' > AA'' = 0$ and $BB'' > BB' = 0$.

Let $T = \{X : X \in \overline{AB} \text{ and } XX' \geq XX''\}$. Let $S = \{AX : X \in T\}$, the set of distances from A covered by the points of T. Since $A \in T$, we have $0 \in S$. Since $T \subseteq \overline{AB}$, S is a subset of $[0, AB]$. Hence, AB is an upper bound for S. By the Least Upper Bound Property, S has a least upper bound c, with $0 \leq c \leq AB$. By Theorem 8.6, there is a point Y on $\overline{AB}$ with $AY = c$. We proceed to show that $YY' = YY''$.

Suppose toward a contradiction that $YY'' < YY'$. By Proposition 21.14(a), there is a point Z in $\overline{AB}$ with Y-Z-B and $ZZ'' < ZZ'$. Then $Z \in T$, and so $AZ \in S$. But $AZ > AY = c$, the least upper bound of S. This is a contradiction.

Suppose toward a contradiction that $YY'' > YY'$. By Proposition 21.14(b), there is a point W in $\overline{AB}$ with A-W-Y, and $WW'' > WW'$. For all points X in $\overline{AB}$ with A-W-X, $XX' < WW'$ and $XX'' > WW''$ (Theorem 21.13). Hence, $XX'' > XX'$. So $X \notin T$. Therefore, $T \subseteq \overline{AW}$ and so $S \subseteq [0, AW]$. Then AW is an upper bound for S. But A-W-Y implies that $c = AY > AW$, and c is the least upper bound for S. This is the contradiction we want.

It is proved that $YY'' \not< YY'$ and $YY'' \not> YY'$. Therefore, $YY'' = YY'$. Since Y' and Y'' are on the same ray with endpoint Y, it follows from Theorem 8.6 that $Y' = Y''$. Thus, C and C_1 meet in the point Y' of the halfplane H. The same argument, applied to the opposite halfplane with edge $\overleftrightarrow{PQ}$, yields a second point where the two circles intersect.

If V is any other point where C and C_1 meet, we may assume that $V \in H$. Now $\triangle PQV \cong \triangle PQY'$ (SSS). Hence, $\angle QPV = \angle QPY'$. Since $\overrightarrow{PV}$ and $\overrightarrow{PY'}$ are both in the fan $\overrightarrow{PQ}\overrightarrow{PV}$ (Theorem 12.2), it follows from Theorem 11.6 that $\overrightarrow{PV} = \overrightarrow{PY'}$. Since $PV = PY'$, Theorem 8.6 implies that $V = Y'$. Therefore, C and C_1 have only two points of intersection.

Theorem 21.16. (Triangle Construction Theorem) Suppose that a, b, c are three positive real numbers such that each is smaller than $\omega/2$, and the sum of any two is larger than the third. Then there exists a triangle whose side lengths are a, b, c.

Proof. We may assume that $a \leq c$ and $b \leq c$. Then $a + b > c$, by hypothesis. There exist points P, Q with $PQ = c$. Let $C = C(P, a)$ and $C_1 = C(Q, b)$. By the Circle Intersection Theorem, C and C_1 meet in a point Y', with P, Q and Y' noncollinear. Then $\triangle PQY'$ has side lengths a, b, c, as desired.

Problem Set 21

1. Prove Proposition 21.1.

2. Prove Proposition 21.2. (Hint: If P is the center of the circle, consider $\overrightarrow{AP}$.)

3. Prove Proposition 21.3.

4. Prove Proposition 21.4. (Hint: Let M be the midpoint of $\overline{AB}$; use Corollary 15.5 to show that $AM < r$. Then apply Proposition 16.9.)

5. Prove that Int $C(P, r)$ is a convex set.

6. Let X be a point in the exterior of a circle C, and suppose that B_1, B_2 are points on C such that $\overleftrightarrow{XB_1}$ and $\overleftrightarrow{XB_2}$ are tangent to C. Prove that $XB_1 = XB_2$.

7. Prove Corollary 21.7. (Hint: What is the distance from P to t?)

8. Prove or find a counterexample: Suppose that $\omega < \infty$ and that a, b, c are positive numbers smaller than ω (but not necessarily all smaller than $\omega/2$) such that the sum of any two is larger than the third. Then there exists a triangle with side lengths a, b, c.

9. Suppose that $C(P, r)$ and $C(Q, t)$ are circles such that $PQ = r + t$. Prove that the circles meet in exactly one point and that they have a common tangent line at this point.

10. Prove Theorem 21.10(b).

11. Prove Theorem 21.11. (Hint: If P is the center of the circle, consider segments PA, PX, PB and relevant triangles.)

12. Prove Theorem 21.12.

The next three problems for this chapter concern an *ellipse*, which is defined as follows: Fix points P_1, P_2 and a number d with $P_1 P_2 < d < \omega$. The *ellipse* $E = E(P_1, P_2, d)$ with *foci*

P_1, P_2 and *aphelion* d is the set of all points each of whose distances from P_1 and P_2 sum to d. That is,

$$E(P_1, P_2, d) = \{X : P_1X + P_2X = d\}.$$

13. Prove that there exist two points X on $\overleftrightarrow{P_1 P_2}$ with $X \in E$. (Hint: Consider $\overrightarrow{P_1 P_2}$ and $\overrightarrow{P_2 P_1}$.)

14. Suppose that A and B are points with $AB < \omega$, $P_1A + P_2A < d$, and $P_1B + P_2B > d$. Prove that $\overline{AB}$ meets E. (Hint: Adapt the proof of Theorem 21.5.)

15. Prove that if A is a point with $P_1A + P_2A < d$ and if h is any ray with endpoint A, then h meets E.

16. Prove the following extension for Euclidean planes of the Inscribed Angle Theorem:

    Let $C(P, r)$ be a circle in a Euclidean plane. Let A and B be distinct points on the circle with $\theta = \angle APB < 180$. Let X be any other point on the circle with $X \notin$ Int $\angle APB$. Then $\angle AXB = \theta/2$. (Hint: Consider separately the cases (i) $\overline{XA}$ is a diameter, (ii) X, B are on the same side of $\overleftrightarrow{PA}$, (iii) $\overline{XB}$ is a diameter, (iv) X, A are on the same side of $\overleftrightarrow{PB}$, (v) $X \in$ Int $\overrightarrow{PA'}\overrightarrow{PB'}$. In case (i), consider $\triangle PXB$. In case (ii), consider $\triangle PXB$ and $\triangle PXA$.)

22   Similarity

The notion of similar geometric figures (which, roughly speaking, have the same shape but not necessarily the same size) is a basic but very powerful idea in Euclidean geometry. Even its most elementary application, the calculation of size of objects where direct measurement is impossible is one of the most widespread uses of mathematics. In this chapter we study the properties of similar triangles. We state and prove the cornerstone of the theory (Theorem 22.10). We then apply it to prove the celebrated theorems of Pythagoras and of Ceva.

Definition. A correspondence $A \leftrightarrow X, B \leftrightarrow Y, C \leftrightarrow Z$ (abbreviated $ABC \leftrightarrow XYZ$) between the vertices of $\triangle ABC$ and those of $\triangle XYZ$ (in an absolute plane) is called a *similarity* if all corresponding angles are congruent. That is, $\angle A = \angle X, \angle B = \angle Y, \angle C = \angle Z$. We denote this by $\triangle ABC \sim \triangle XYZ$, and say that $\triangle ABC$ is *similar* to $\triangle XYZ$ *under the correspondence* $ABC \leftrightarrow XYZ$ (Figure 22.1).

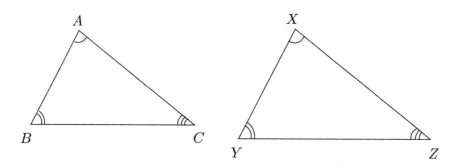

Figure 22.1

Of course, if two triangles are congruent, then they are similar.

What is perhaps surprising is that in the non-Euclidean absolute planes, the converse of this result is true. That is, if $\mathbb{P}$ is a hyperbolic or spherical absolute plane, and if two triangles in $\mathbb{P}$ are similar as in the preceding definition, then the triangles are congruent. The reason why this holds, as we shall soon see, is that in a non-Euclidean absolute plane the angle sum of a triangle is never 180 (Theorem 19.4).

Definition.   Let $\triangle ABC$ be a triangle in an absolute plane. The number $\delta(ABC) = 180 - \sigma(ABC)$ is called the *defect* of $\triangle ABC$.

The next observation just restates part of Theorem 19.4 in terms of this notation.

Proposition 22.1.   Let $\triangle ABC$ be a triangle in an absolute plane $\mathbb{P}$. Then $\delta(ABC) = 0$ when $\mathbb{P}$ is Euclidean, $\delta(ABC) < 0$ when $\mathbb{P}$ is spherical, and $0 < \delta(ABC) < 180$ when $\mathbb{P}$ is hyperbolic.

Proposition 22.2.   If $\triangle ABC$ is a triangle in an absolute plane, and if B-D-C and A-Q-C, then $\delta(ABC) = \delta(ABD) + \delta(ACD)$ and $\delta(ABC) = \delta(ABD) + \delta(ADQ) + \delta(CDQ)$.

Proof.   Problem 2.

The diligent reader will recall that the definition of defect appeared previously in Problem 2 of Chapter 19, and already will have met the next theorem as Problem 4 in Chapter 19.

Theorem 22.3.   (AAA Criterion for Congruence).   Suppose that $\triangle ABC$ and $\triangle XYZ$ are triangles in a non-Euclidean absolute plane with $\angle A = \angle X$, $\angle B = \angle Y$, and $\angle C = \angle Z$. Then $\triangle ABC \cong \triangle XYZ$.

Proof.   If any of $AB = XY$, $BC = YZ$, or $CA = ZX$ is true, then the result follows by ASA. So it suffices to assume that all of $AB - XY$, $BC - YZ, CA - ZX$ are nonzero and to work toward a contradiction. Under this assumption, at least two of these numbers must have the same sign. We may switch the notation for $\triangle ABC$ and $\triangle XYZ$ if necessary, so there is no loss of generality to assume that $AB > XY$ and $CA > ZX$. Then there exist points P on $\overline{AB}$ and Q on $\overline{AC}$ so that $AP = XY$ and $AQ = XZ$ (8.6). Hence, $\triangle APQ \cong \triangle XYZ$ by SAS (Figure 22.2).

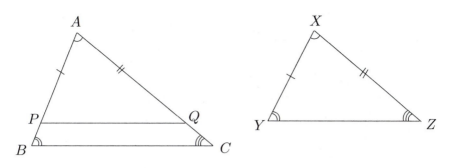

Figure 22.2

By hypothesis, $\sigma(ABC) = \sigma(XYZ)$ and so $\delta(ABC) = \delta(XYZ)$. Since $\triangle APQ \cong \triangle XYZ$, we also have $\delta(APQ) = \delta(XYZ)$. Now an easy argument based on Propositions 22.2 and 22.1 shows that $\delta(BPQ) = \delta(QBC) = 0$. Therefore, the plane is Euclidean by Proposition 22.1 and Theorem 19.4. This contradicts our hypothesis.

We assume for the rest of this chapter that $\mathbb{P}$ is a Euclidean plane. The next result shows that under this hypothesis, there is some redundancy in the definition of similar triangles.

Proposition 22.4. Suppose that $\triangle ABC$ and $\triangle XYZ$ are triangles (in a Euclidean plane!) such that $\angle A = \angle X$ and $\angle B = \angle Y$. Then $\triangle ABC \sim \triangle XYZ$.

Proof. Problem 4.

We need to generalize slightly the notion of transversal configuration from Chapter 17, in order to cover the situation where a given line meets several (or many) other given lines in distinct points.

Definition. A *transversal configuration* is an ordered $(k+1)$-tuple $(t, m_1, m_2, \cdots, m_k)$ of $k+1$ different lines so that t meets the lines $m_1, m_2, \cdots, m_k$ in distinct points $M_1, M_2, \cdots, M_k$, respectively. Line t is called the *transversal* (Figure 22.3).

We will always label lines $m_1, \cdots, m_k$ so that M_1-M_2-M_3, M_2-M_3-M_4, $\cdots$, M_{k-2}-M_{k-1}-M_k.

Proposition 22.5. Assume that $m \parallel n$. (a) If t is a line different from m that meets m, then t also meets n.

(b) All points of n lie in the same halfplane with edge m.

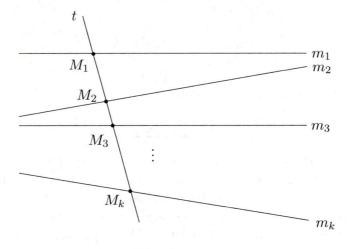

Figure 22.3

Proof. (a) If $t \parallel n$, then m and t would be two lines through the same point and parallel to n. This contradicts the standing assumption that the plane is Euclidean.

(b) The proof of (b) is Problem 4 in Chapter 17.

The next result was part of Problem 9 in Chapter 17. We include a proof here.

Theorem 22.6. Suppose that $t \parallel l$, $m \parallel n$, t meets m, n in points A, C, respectively, and l meets m, n in points B, D, respectively (Figure 22.4). Then $AB = CD$ and $AC = BD$.

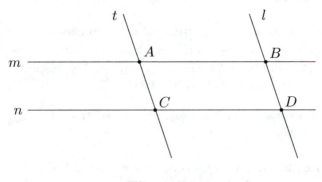

Figure 22.4

Proof. Given any three of the points A, B, C, D, two of the three are on one line and the third is on a parallel line; so the three points are noncollinear. In particular, triangles $\triangle ABD$ and $\triangle DCA$ are defined.

Now $(\overleftrightarrow{AD}, m, n)$ and $(\overleftrightarrow{AD}, t, l)$ are transversal configurations with $m \parallel n$ and $t \parallel l$. By Theorem 17.7, $\angle BAD = \angle CDA$ and $\angle CAD = \angle BDA$ (alternate interior angles are congruent). Then $\triangle ABD \cong \triangle DCA$ by ASA. Therefore, $AB = DC$ and $BD = CA$.

Theorem 22.7. (Congruent Segments Theorem) Suppose that l, m, n are three distinct parallel lines (with l, n on opposite sides of m), and t is a transversal that meets l, m, n in points A, B, C, respectively, such that $\overline{AB} \cong \overline{BC}$. Suppose that s is any other transversal that meets l, m, n in points D, E, F, respectively. Then $\overline{DE} \cong \overline{EF}$.

Proof. Assume first that $B \neq E$. Since $\mathbb{P}$ is Euclidean, there is a unique line p through E with $p \parallel t$. Then p meets l, n in points G, H, respectively by Proposition 22.5(a) (Figure 22.5).

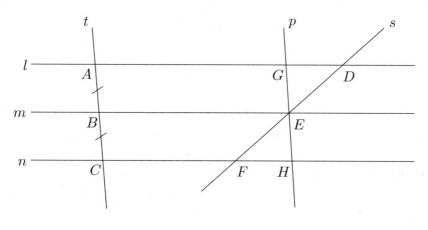

Figure 22.5

Now $GE = AB$ and $EH = BC$ by Theorem 22.6. Since $AB = BC$ by hypothesis, we have $GE = EH$. So if $s = p$, we are done; thus we may assume $p \neq s$. Then $D \neq G$ and $F \neq H$. Note that G-E-H and hence G and H are on opposite sides of s. Likewise, D and F are on opposite sides of p.

Theorem 17.7 applied to (p, l, n) yields $\angle DGE = \angle FHE$. Also, $\angle GED = \angle HEF$ by Theorem 14.2. So $\triangle GED \cong \triangle HEF$ by ASA. Thus, $DE = FE$ and the theorem is proved if $B \neq E$. If $B = E$, then the same argument, with t in place of p, completes the proof.

Note that the Congruent Segments Theorem is just the particular case of the next result where $AB/BC = 1$.

Theorem 22.8. (Proportional Segments Theorem) Suppose that l, m, n are three distinct parallel lines (with l, n on opposite sides of m), and t, s are two transversals so that t meets l, m, n in points A, B, C, respectively, and s meets l, m, n in points D, E, F, respectively. Then $AB/BC = DE/EF$ (Figure 22.6).

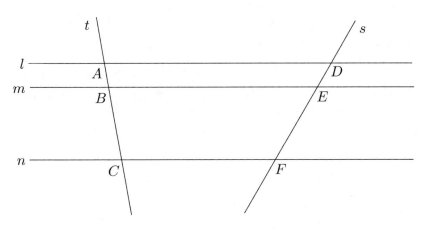

Figure 22.6

Proof. We first consider the case where AB/BC is a rational number; that is, $AB/BC = p/q$ for some positive integers p and q. By repeated use of Theorem 8.6, we may divide $\overline{BC}$ into q congruent segments by selecting points $P_1, P_2, \cdots, P_{q-1}$ so that B-P_1-C, P_1-P_2-C, P_2-P_3-C, $\cdots P_{q-2}$-P_{q-1}-C and $BP_1 = P_1P_2 = P_2P_3 = \cdots = P_{q-2}P_{q-1} = P_{q-1}C = \dfrac{1}{q}BC$.

Now $AB = (p/q)BC = p(\dfrac{1}{q}BC)$. We may divide $\overline{AB}$ into p congruent segments, since by Theorem 8.6 there exist points $Q_1, Q_2, \cdots, Q_{p-1}$ so that A-Q_1-B, Q_1-Q_2-B, Q_2-Q_3-B, $\cdots$, Q_{p-2}-Q_{p-1}-B and $AQ_1 = Q_1Q_2 = Q_2Q_3 = \cdots = Q_{p-2}Q_{p-1} = Q_{p-1}B = \dfrac{1}{p}AB = \dfrac{1}{q}BC$.

Let $l_1, l_2, \cdots, l_{p-1}$ be the lines through $Q_1, Q_2, \cdots, Q_{p-1}$, respectively, that are parallel to l (hence to m and n by Proposition 22.5(a)), and let $m_1, m_2, \cdots, m_{q-1}$ be the lines through $P_1, P_2, \cdots, P_{q-1}$, respectively, that are parallel to l, m, n. Then (again by Proposition 22.5(a)) s meets $l_1, l_2, \cdots, l_{p-1}, m_1, m_2, \cdots, m_{q-1}$ in points $R_1, R_2, \cdots, R_{p-1}, S_1, S_2, \cdots, S_{q-1}$, respectively (Figure 22.7).

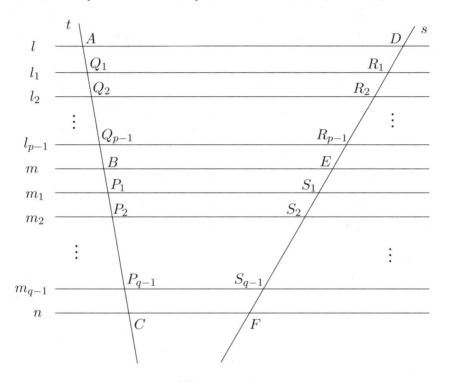

Figure 22.7

Since l and m are on opposite sides of l_1, it follows that D-R_1-E; and similarly R_1-R_2-E, R_2-R_3-E, $\cdots$, R_{p-2}-R_{p-1}-E, E-S_1-F, S_1-S_2-F, S_2-S_3-F, $\cdots$, S_{q-2}-S_{q-1}-F. Since $AQ_1 = Q_1Q_2$, the Congruent Segments Theorem implies that $DR_1 = R_1R_2$. Similarly, $Q_1Q_2 = Q_2Q_3$ yields that $R_1R_2 = R_2R_3$. Continuing, we obtain

$$DR_1 = R_1R_2 = R_2R_3 = \cdots = R_{p-2}R_{p-1} = R_{p-1}E$$
$$= ES_1 = S_1S_2 = \cdots = S_{q-2}S_{q-1} = S_{q-1}F.$$

The equal lengths $DR_1, R_1R_2, \cdots, R_{p-1}E$ sum to DE; hence $DE = pDR_1$. Likewise, $ES_1, S_1S_2, \cdots, S_{q-1}F$ sum to EF, so $EF = qES_1 = qDR_1$. Therefore,

$$DE/EF = p(DR_1)/q(DR_1) = p/q = AB/BC.$$

Next we treat the case $AB/BC = z$, where z is an irrational positive real number. We shall use (without proving it) the well-known fact that z can be approximated arbitrarily closely by rational numbers x and y with $x < z < y$.

Fix any positive rational numbers x and y with $x < z = (AB/BC) < y$. Then $xBC < zBC = AB < yBC$. So Theorem 8.6 implies that there is a point X on $\overline{BA}$ with $BX = xBC$; and there is a point Y on $\overrightarrow{BA}$ but not on $\overline{BA}$ with $BY = yBC$. Let u be the line through X with $u \parallel l$, and let v be the line through Y with $v \parallel l$. Then l and m are on opposite sides of u, and m and v are on opposite sides of l. So u meets s in point G where E-G-D, and v meets s in point H with E-D-H (Figure 22.8).

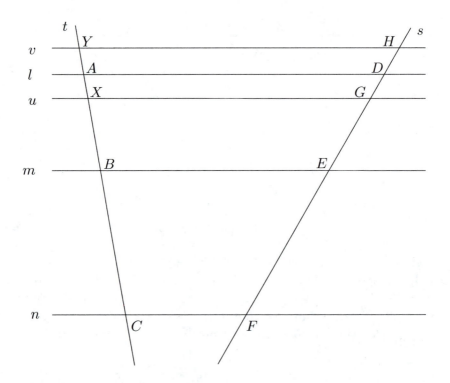

Figure 22.8

Since $XB/BC = x$ and $YB/BC = y$ are rational, what we have proved so far shows that

$$x = XB/BC = GE/EF \text{ and } y = YB/BC = HE/EF.$$

Since $GE < DE < HE$, we have

$$x = GE/EF < DE/EF < HE/EF = y.$$

Therefore, DE/EF lies between all pairs of rational numbers x, y such that $x < z < y$. Since such x and y can be chosen arbitrarily close to z, it follows that $DE/EF = z = AB/BC$.

Corollary 22.9. Suppose that l, m, n are three distinct parallel lines (with l, n on opposite sides of m), and t, s are two transversals so that t meets l, m, n in points A, B, C, respectively, and s meets l, m, n in points D, E, F, respectively. Then $AC/AB = DF/DE$.

Proof. Problem 5.

Next we prove the fundamental result that in a Euclidean plane, two triangles are similar if and only if their corresponding sides are proportional.

Theorem 22.10. (Similarity Theorem) Let $\triangle ABC$ and $\triangle XYZ$ be triangles in $\mathbb{P}$ (still assumed Euclidean). Then the following statements are equivalent:

(a) $\triangle ABC \sim \triangle XYZ$;

(b) $AB/XY = BC/YZ = CA/ZX$;

(c) there is some number $k > 0$ with $AB = kXY$, $BC = kYZ$, $CA = kZX$;

(d) $AB/BC = XY/YZ, AB/CA = XY/ZX$, and $BC/CA = YZ/ZX$.

Proof. We first show that (a) $\Rightarrow$ (b). Assume that $\triangle ABC \sim \triangle XYZ$. If $AB = XY$, then $\triangle ABC \cong \triangle XYZ$ by ASA, and so (b) holds with each ratio equal to 1. So we may assume that $AB \neq XY$. Since we may switch the notation for the two triangles if necessary, there is no loss of generality in assuming that $AB < XY$. Then there is a point U in $\overline{XY}$ with $XU = AB$ (Theorem 8.6). Let l be the line through U with $l \parallel \overleftrightarrow{YZ}$, and let n be the line through X with $n \parallel \overleftrightarrow{YZ}$.

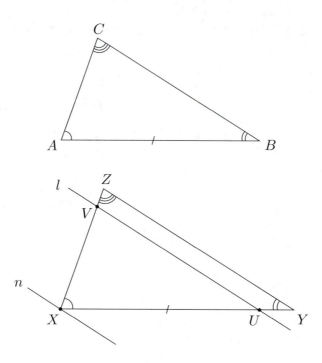

Figure 22.9

Since Y and X, and hence Z and X, are on opposite sides of l, l meets $\overline{XZ}$ at some point V (Axiom S) (Figure 22.9). Now Theorem 17.7 applied to $(\overleftrightarrow{XY}, l, \overleftrightarrow{YZ})$ and Theorem 14.2 yield that $\angle ZYU = \angle VUX$. But $\angle ZYU = \angle ZYX = \angle Y = \angle B$ by (a) and the definition of similar triangles. Since $XU = AB$ and $\angle X = \angle A$ (again by (a)), $\triangle ABC \cong \triangle XUV$ by ASA. So $AC = XV$.

Corollary 22.9, applied to the parallel lines $n, l, \overleftrightarrow{YZ}$ with transversals $\overleftrightarrow{XZ}$ and $\overleftrightarrow{XY}$, yields that $XV/XZ = XU/XY$. Hence, $AC/XZ = AB/XY$.

Let W be the point in $\overline{XY}$ with $YW = BA$ (Theorem 8.6). Let t be the line through W with $t \parallel \overleftrightarrow{XZ}$, and let m be the line through Y with $m \parallel \overleftrightarrow{XZ}$. Then an argument similar to the previous one, with t and m in place of l and n, implies that $BC/YZ = AB/XY$. So $AC/XZ = AB/XY = BC/YZ$, and thus (a) implies (b).

Next we show that (b) $\Rightarrow$ (a). Assume that $AB/XY = BC/YZ = CA/XZ$. If each of these ratios is 1, then $\triangle ABC \cong \triangle XYZ$ by

SSS, and (a) holds. So we may assume that the ratios do not equal 1. Switching the roles of the two triangles if necessary, we may suppose without loss of generality that the ratios are larger than 1. Hence, $AB > XY$. Then there is a point P in $\overline{AB}$ with $AP = XY$. Let l, m be the lines through A, P, respectively, that are parallel to $\overleftrightarrow{BC}$ (Figure 22.10).

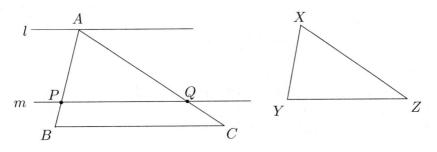

Figure 22.10

Now A, B on opposite sides of m implies that A, C are also on opposite sides of m. Thus, $\overline{AC}$ meets m at a point Q. Corollary 22.9, applied to transversals $\overleftrightarrow{AB}$ and $\overleftrightarrow{AC}$ across parallels l, m and $\overleftrightarrow{BC}$, yields $AB/AP = AC/AQ$. But $AP = XY$, and $AB/XY = AC/XZ$, by (b). Thus, $AC/AQ = AC/XZ$, and so $AQ = XZ$.

Let p, n be the lines through C, Q, respectively, that are parallel to $\overleftrightarrow{AB}$. Then as previously, n meets $\overline{BC}$ at a point R (Figure 22.11).

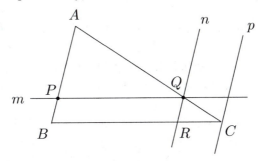

Figure 22.11

By Corollary 22.9, $AC/AQ = BC/BR$. But $AC/AQ = AC/XZ = BC/YZ$, from (b). Therefore, $YZ = BR$. Now $BR = PQ$ by Theorem 22.6. We have $AP = XY, AQ = XZ$ and $PQ = YZ$.

Then $\triangle APQ \cong \triangle XYZ$ by SSS. So $\angle A = \angle PAQ = \angle X, \angle APQ = \angle Y$ and $\angle AQP = \angle Z$. By Theorems 14.2 and 17.7 applied to $(\overleftrightarrow{AB}, m, \overleftrightarrow{BC})$, $\angle B = \angle APQ$. It follows from $(\overleftrightarrow{AC}, m, \overleftrightarrow{BC})$ that $\angle C = \angle AQP$. We have established that $\triangle ABC \sim \triangle XYZ$, so (a) holds.

It is an easy exercise in arithmetic to show that each of (c) and (d) is equivalent to (b).

There are hundreds of known proofs of the famous Pythagorean Theorem on the sides of a right triangle. The following proof is based on the Similarity Theorem.

Theorem 22.11. (Pythagoras) Let $\triangle ABC$ be a right triangle in a Euclidean plane with right angle $\angle C$. Let $AB = c, BC = a, CA = b$. Then $a^2 + b^2 = c^2$ (Figure 22.12).

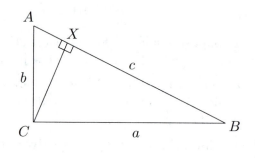

Figure 22.12

Proof. Let X be the closest point on $\overleftrightarrow{AB}$ to C. Then $\overleftrightarrow{CX} \perp \overleftrightarrow{AB}$ by Theorem 16.8. Note that $\angle CBA$ and $\angle CAB$ are acute (Theorem 16.4). So by the Acute Angle Theorem (Theorem 20.2), X is in $\overrightarrow{BA} \cap \overrightarrow{AB}$ and hence A-X-B. Then $\triangle ACX$ and $\triangle CBX$ are right triangles with $\angle CAB = \angle CAX$ and $\angle CBA = \angle CBX$. It is left as part of Problem 7 to show that $\triangle ACX \sim \triangle ABC$ and $\triangle ABC \sim \triangle CBX$ (the ordering of the vertices is important). Then it follows from the Similarity Theorem that $AX/AC = AC/AB$ and $BX/BC = BC/BA$. These equations imply that $a^2 + b^2 = c^2$. The details are the rest of Problem 7.

The following remarkable result of Giovanni Ceva is relatively recent on the time line of geometry, having been published in 1678. But it is quite similar to a theorem of Menelaus that was proved about 1600 years earlier. Ceva's theorem determines precisely when three Cevian lines, chosen generally via the three vertices and opposite sides of a given triangle, will be concurrent.

Theorem 22.12. (Ceva) Let $\triangle ABC$ be a triangle in a Euclidean plane, and let U, V, W be points with $U \in \overline{BC}^o, V \in \overline{AC}^o, W \in \overline{AB}^o$. Then the three lines $\overleftrightarrow{AU}, \overleftrightarrow{BV}, \overleftrightarrow{CW}$ are concurrent if and only if $\dfrac{AV}{VC} \cdot \dfrac{CU}{UB} \cdot \dfrac{BW}{WA} = 1$.

Proof. Suppose first that $\overleftrightarrow{AU}, \overleftrightarrow{BV}, \overleftrightarrow{CW}$ meet in a point P. Then A-P-U, B-P-V, C-P-W (20.3). Let l be the unique line through A with $l \parallel \overleftrightarrow{BC}$. Then $\overleftrightarrow{CW}$ and $\overleftrightarrow{BV}$ meet l at some points M, N, respectively (Figure 22.13).

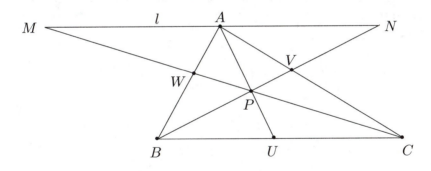

Figure 22.13

Now W and A are in the same halfplane with edge $\overleftrightarrow{BC}$ (Theorem 10.3). Hence, Proposition 22.5(b) implies that line l lies in this half-plane; therefore, so does M. Then by Theorem 10.3 again, M is on $\overrightarrow{CW}$ and not on the opposite ray $\overrightarrow{CW}'$. If C-M-W, then $\overrightarrow{AC}$-$\overrightarrow{AM}$-$\overrightarrow{AW}$ (Axiom C). So $\overrightarrow{AM}$ would meet $\overline{BC}$, by the Crossbar Theorem. This contradicts $l \parallel \overleftrightarrow{BC}$. Therefore, C-W-M. Similarly, B-V-N. Now it is clear that M and N are on opposite sides of $\overleftrightarrow{AB}$. So M-A-N.

Each of the following four pairs of triangles is similar (details are left for Problem 8): $\triangle BVC \sim \triangle NVA, \triangle PCU \sim \triangle PMA, \triangle PBU \sim \triangle PNA, \triangle BWC \sim \triangle AWM$. The Similarity Theorem, applied to each of these relations in turn, yields the following four equalities:

$$\frac{AV}{CV} = \frac{AN}{CB}, \quad \frac{CU}{PU} = \frac{MA}{PA}, \quad \frac{PU}{UB} = \frac{PA}{AN}, \quad \frac{BW}{AW} = \frac{BC}{AM}.$$

Therefore,

$$\frac{AV}{VC} \cdot \frac{CU}{UB} \cdot \frac{BW}{WA} = \frac{AV}{CV} \cdot \frac{CU}{PU} \cdot \frac{PU}{UB} \cdot \frac{BW}{AW}$$

$$= \frac{AN}{CB} \cdot \frac{MA}{PA} \cdot \frac{PA}{AN} \cdot \frac{BC}{AM} = 1.$$

To prove the converse, suppose that $\dfrac{AV}{VC} \cdot \dfrac{CU}{UB} \cdot \dfrac{BW}{WA} = 1$. Now $\overleftrightarrow{CW}$ and $\overleftrightarrow{BV}$ meet at a point P with C-P-W and B-P-V (20.3). By Axiom C and the Crossbar Theorem, $\overrightarrow{AP}$ meets $\overline{BC}^o$ at some point X (Figure 22.14).

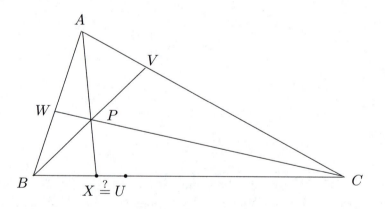

Figure 22.14

By what was proved previously and by hypothesis,

$$\frac{AV}{VC} \cdot \frac{CX}{XB} \cdot \frac{BW}{WA} = 1 = \frac{AV}{VC} \cdot \frac{CU}{UB} \cdot \frac{BW}{WA}.$$

Therefore, $\dfrac{CX}{XB} = \dfrac{CU}{UB}$, so $CX = \dfrac{CU}{UB} \cdot XB$. Since C-X-B and C-U-B,

$$\left(\frac{CU}{UB}+1\right)\cdot XB \;=\; \left(\frac{CU}{UB}\cdot XB\right)+XB = CX+XB = CB$$

$$=\; CU+UB = \left(\frac{CU}{UB}+1\right)\cdot UB.$$

Therefore, $XB = UB$. By Theorem 8.6 applied to $\overline{BC}, X = U$. Hence, $\overleftrightarrow{AU}, \overleftrightarrow{BV}, \overleftrightarrow{CW}$ are indeed concurrent at P.

Corollary 22.13. The three medians of any triangle in a Euclidean plane are concurrent.

Proof.   Let $\overleftrightarrow{AU}, \overleftrightarrow{BV}, \overleftrightarrow{CW}$ be the medians of $\triangle ABC$, with $U \in \overline{BC}^{o}, V \in \overline{AC}^{o}, W \in \overline{AB}^{o}$. Then by definition of "median," $BU = UC, AV = VC, AW = WB$. Then it is immediate that $\dfrac{AV}{VC}\cdot\dfrac{CU}{UB}\cdot\dfrac{BW}{WA} = 1$. So Ceva's theorem implies the result.

Problem Set 22

A Euclidean plane is the context for each problem.

1. Given $\triangle ABC$, let P, Q be points with A-P-B, A-Q-C, and $\angle AQB = \angle ABC$. Find a formula for AP in terms of PB, AQ, and AC. (Hint: Find a relevant pair of equal ratios; then look for a quadratic equation in AP.)

2. Prove Proposition 22.2.

3. Fill in the details of the proof of Theorem 22.3. Specifically, verify that $\delta(BPQ) = \delta(QBC) = 0$.

4. Prove Proposition 22.4.

5. Prove Corollary 22.9.

6. Show that each of the statements (c), (d) of Theorem 22.10 is equivalent to statement (b).

7. Complete the proof of Theorem 22.11. That is, show that $\triangle ACX \sim \triangle ABC, \triangle ABC \sim \triangle CBX$, and then that $a^2 + b^2 = c^2$.

8. Prove that the claimed similarity of each of the four pairs of triangles in the proof of Theorem 22.12 is indeed true.

9. Given $\triangle ABC$, let M and N be the midpoints of sides $\overline{AB}$ and $\overline{AC}$, respectively. Prove that $BC = 2MN$.

10. Suppose that $l \parallel n$ and that t, s are two transversals such that t meets l, n in points A, C, respectively, and s meets l, n in points D, F, respectively. Suppose also that B, E are points such that $A\text{-}B\text{-}C$, $D\text{-}E\text{-}F$ and $AB/AC = DE/DF$. Prove that $\overleftrightarrow{BE} \parallel l$. (This is the converse of Corollary 22.9.) (Hint: What line through B might you consider?)

11. Suppose that $\triangle ABC$ and $\triangle XYZ$ are such that $\angle A = \angle X$ and $AB/XY = AC/XZ$. Prove that $\triangle ABC \sim \triangle XYZ$. (This is the SAS criterion for similarity). (Hint: Study the beginning of the argument for (b) $\Rightarrow$ (a) in the proof of Theorem 22.10 and use Problem 10.)

12. Prove the converse of the Pythagorean theorem: Let $\triangle ABC$ be a triangle with $c = AB, a = BC, b = CA$. If $c^2 = a^2 + b^2$, then $\triangle ABC$ is a right triangle with hypotenuse $\overline{AB}$.

13. The following statement is named for the English mathematician John Wallis (1616–1703), who proposed it as a substitute for Euclid's Fifth Postulate:

Wallis's Postulate: Given any triangle $\triangle ABC$ and any segment $\overline{DE}$ in $\mathbb{P}$, then there exists a point F in $\mathbb{P}$ such that $\triangle DEF \sim \triangle ABC$.

(a) Assume that $\mathbb{P}$ is a Euclidean plane and prove that the statement of Wallis's postulate holds (that is, it's a theorem for Euclidean planes).

(b) Assume that $\mathbb{P}$ is an absolute plane in which Wallis's postulate is true. Prove that $\mathbb{P}$ must be Euclidean.

14. Prove Ptolemy's Theorem for an inscribed quadilateral, the statement of which is as follows:

Let A, B, C, D be distinct points on a circle in a Euclidean plane such that $\overline{AB} \cap \overline{CD} = \emptyset$, $\overline{BC} \cap \overline{AD} = \emptyset$ and $\overline{BD} \cap \overline{AC} \neq \emptyset$ (Figure 22.15). Then

$$(AB)(CD) + (AD)(BC) = (AC)(BD).$$

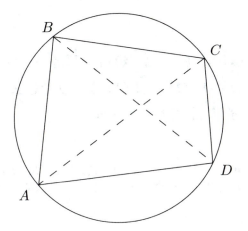

Figure 22.15

(Hint: We may assume by choice of notation that $\angle ABD \geq \angle CBD$. Show by Theorem 11.6 and Theorem 12.4 that there is a point E on $\overline{AC}$ such that $\angle ABE = \angle CBD$. Use Problem 16 in Chapter 21 to show that $\triangle BAE \sim \triangle BDC$ and $\triangle ABD \sim \triangle EBC$.)

Appendix I. Definitions and Assumptions from Book I of Euclid's Elements

These are taken directly from T. L. Heath, *The Thirteen Books of Euclid's Elements, Vol. I*, Dover Publications, New York, 1956, pp. 153–155.

Definitions

1. A *point* is that which has no part.

2. A *line* is breadthless length.

3. The extremities of a line are points.

4. A *straight line* is a line which lies evenly with the points on itself.

5. A *surface* is that which has length and breadth only.

6. The extremities of a surface are lines.

7. A *plane surface* is a surface which lies evenly with the straight lines on itself.

8. A *plane angle* is the inclination to one another of two lines in a plane which meet one another and do not lie in a straight line.

9. And when the lines containing the angles are straight, the angle is called *rectilineal*.

10. When a straight line set up on a straight line makes the adjacent angles equal to one another, each of the equal angles is *right*, and the straight line standing on the other is called a *perpendicular* to that on which it stands.

11. An *obtuse* angle is an angle greater than a right angle.

12. An *acute* angle is an angle less than a right angle.

13. A *boundary* is that which is an extremity of anything.

14. A *figure* is that which is contained by any boundary or boundaries.

15. A *circle* is a plane figure contained by one line such that all the straight lines falling upon it from one point among those lying within the figure are equal to one another.

16. And the point is called the *center* of the circle.

17. A *diameter* of the circle is any straight line drawn through the center and terminated in both directions by the circumference of the circle, and such a straight line also bisects the circle.

18. A *semicircle* is the figure contained by the diameter and the circumference cut off by it. And the center of the semicircle is the same as that of the circle.

19. *Rectilineal figures* are those which are contained by straight lines, *trilateral* figures being those contained by three, *quadrilateral* those contained by four, and *multilateral* those contained by more than four straight lines.

20. Of trilateral figures, an *equilateral triangle* is that which has three sides equal, an *isosceles triangle* that which has two of its sides alone equal, and a *scalene triangle* that which has its three sides unequal.

21. Further, of trilateral figures, a *right-angled triangle* is that which has a right angle, an *obtuse-angled triangle* that which has an obtuse angle, and an *acute-angled triangle* that which has its three angles acute.

22. Of quadrilateral figures, a *square* is that which is both equilateral and right-angled; an *oblong* that which is right-angled but not equilateral; a *rhombus* that which is equilateral but not right-angled; and a *rhomboid* that which has its opposite sides and angles equal to one another but is neither equilateral nor right-angled. And let quadrilaterals other than these be called *trapezia*.

23. *Parallel* straight lines are straight lines which, being in the same plane and being produced indefinitely in both directions, do not meet one another in either direction.

Postulates

Let the following be postulated:

1. To draw a straight line from any point to any point.

2. To produce a finite straight line continuously in a straight line.

3. To describe a circle with any center and distance.

4. That all right angles are equal to one another.

5. That, if a straight line falling on two straight lines makes the interior angles on the same side less than two right angles, the two straight lines, if produced indefinitely, meet on that side on which are the angles less than the two right angles.

Common Notions (Axioms)

1. Things which are equal to the same thing are also equal to one another.

2. If equals be added to equals, the wholes are equal.

3. If equals be subtracted from equals, the remainders are equal.

4. Things which coincide with one another are equal to one another.

5. The whole is greater than the part.

Appendix II. Side-Angle-Side and the Hyperbolic Plane

We show here that the Side-Angle-Side criterion for congruence of triangles is valid in our model $\mathbb{H}$ of the hyperbolic plane. This fact was claimed without proof in Chapter 13 and is certainly not obvious. We assume as in the text that $\mathbb{H}$ satisfies all of the axioms through Chapter 11 and hence all of the theorems through Chapter 12.

First we derive a formula for hyperbolic angle measure as defined in Chapter 11. We presuppose two well-known results, the second from elementary calculus: (1) The *Law of Cosines* for Euclidean triangles: If γ is the measure of $\angle C$ in $\triangle ABC$ (in a Euclidean plane) and if $a = BC$, $b = AC$ and $c = AB$, then

$$c^2 = a^2 + b^2 - 2ab\cos\gamma.$$

(2) If a curve in $\mathbb{R}^3$ is parametrized as $(f(t), g(t), h(t))$ for $t \in \mathbb{R}$, then the tangent vector to the curve at point $(f(t_0), g(t_0), h(t_0))$ is $(f'(t_0), g'(t_0), h'(t_0))$ (where f' denotes the derivative of f, etc.). That is, $(f'(t_0),\ g'(t_0),\ h'(t_0))$ are the coordinates of the head of the vector when its tail is placed at the origin. When the tail is at $(f(t_0), g(t_0), h(t_0))$, then of course the head is at $(f(t_0)+f'(t_0), g(t_0)+g'(t_0), h(t_0) + h'(t_0))$.

Proposition A1. Let $P = (x_0, y_0)$, $Q = (x_1, y_1)$, $R = (x_2, y_2)$ be noncollinear points in $\mathbb{H}$. Assume that $x_0 < x_1$ and that $\overleftrightarrow{PQ}$ has the equation $y = mx + b$.

(a) If $\overleftrightarrow{PR}$ has the equation $y = nx + c$ and $x_0 < x_2$, then

265

(ℍ-measure) $\angle QPR = \cos^{-1}\left(\dfrac{1 + mn - bc}{\sqrt{1 + m^2 - b^2}\sqrt{1 + n^2 - c^2}}\right);$

(b) if $\overleftrightarrow{PR}$ is the vertical line $x = x_0$ and if $y_0 < y_2$ then

(ℍ-measure) $\angle QPR = \cos^{-1}\left(\dfrac{m + bx_0}{\sqrt{1 + m^2 - b^2}\sqrt{1 - x_0^2}}\right);$

where the values of $\cos^{-1}$ are given in degrees in both cases.

Proof.   The hyperbolic measure of $\angle QPR$ was defined in Chapter 11 as follows: Project rays $\overrightarrow{PQ}$ and $\overrightarrow{PR}$ vertically upward onto the upper unit hemisphere; so the images are circular arcs. The tangent vectors to these arcs (at the point to which P projects) form an angle whose Euclidean measure is defined to be the (ℍ-measure) $\angle QPR$. (See Figure 11.9.) Since $x_0 < x_1$, the circular arc above $\overrightarrow{PQ}$ is parametrized as

$$\left(x, mx + b, \sqrt{1 - x^2 - (mx + b)^2}\right),$$
$$\text{for } x_0 \le x < \left(-mb + \sqrt{1 + m^2 - b^2}\right)\big/(1 + m^2).$$

Hence, the desired tangent vector to this arc is

$$\left(\frac{dx}{dx}, \frac{d}{dx}(mx + b), \frac{d}{dx}\left(\sqrt{1 - x^2 - (mx + b)^2}\right)\right) \text{ (at } x = x_0)$$
$$= \left(1, m, \frac{-x_0 - m(mx_0 + b)}{(1 - x_0^2 - (mx_0 + b)^2)^{1/2}}\right)$$
$$= (1, m, -(x_0 + my_0)(1 - x_0^2 - y_0^2)^{-1/2}).$$

If the hypothesis of (a) holds, then the tangent vector at the projection of P to the arc above $\overrightarrow{PR}$ has the similar form $(1, n, -(x_0 + ny_0)(1 - x_0^2 - y_0^2)^{-1/2})$. Then (ℍ-measure) $\angle QPR$ equals the Euclidean measure γ of $\angle C$ in $\triangle ABC$, where $C = (0, 0, 0)$,

$$A = (1, m, -(x_0 + my_0)(1 - x_0^2 - y_0^2)^{-1/2}),$$
$$B = (1, n, -(x_0 + ny_0)(1 - x_0^2 - y_0^2)^{-1/2}).$$

By the Law of Cosines,

$$\cos\gamma = \frac{(AC)^2 + (BC)^2 - (AB)^2}{2(AC)(BC)}.$$

Since

$$AC = \left(1 + m^2 + \frac{(x_0 + y_0 m)^2}{1 - x_0^2 - y_0^2}\right)^{1/2},$$

$$BC = \left(1 + n^2 + \frac{(x_0 + y_0 n)^2}{1 - x_0^2 - y_0^2}\right)^{1/2}, \quad \text{and}$$

$$(AB)^2 = (m - n)^2 - \frac{(m - n)^2 y_0^2}{1 - x_0^2 - y_0^2},$$

algebraic calculation (left as an exercise) yields that $\cos\gamma$ equals

$$\frac{1 + mn - (y_0 - mx_0)(y_0 - nx_0)}{(1 + m^2 - (y_0 - mx_0)^2)^{1/2}(1 + n^2 - (y_0 - nx_0)^2)^{1/2}}$$
$$= \frac{1 + mn - bc}{\sqrt{1 + m^2 - b^2}\sqrt{1 + n^2 - c^2}}.$$

So the conclusion of (a) follows.

If the hypothesis of (b) holds for $\overrightarrow{PR}$, then the circular arc above $\overrightarrow{PR}$ is parametrized as

$$\left(x_0, y, \sqrt{1 - x_0^2 - y^2}\right), \quad \text{for } y_0 \le y \le \sqrt{1 - x_0^2}.$$

Thus the tangent vector to this arc at the projection of point P is

$$\left(\frac{d}{dy}(x_0), \frac{d}{dy}(y), \frac{d}{dy}\left(\sqrt{1 - x_0^2 - y^2}\right)\right) \quad \text{(evaluated at } y = y_0)$$
$$= \left(0, 1, -y_0(1 - x_0^2 - y_0^2)^{-1/2}\right).$$

Let

$$A = (1, m, -(x_0 + y_0 m)(1 - x_0^2 - y_0^2)^{-1/2}),$$
$$B = (0, 1, -y_0(1 - x_0^2 - y_0^2)^{-1/2}),$$

$C = (0, 0, 0)$. Then the Law of Cosines for $\triangle ABC$, with γ as the Euclidean measure of $\angle C$, yields

$$\cos\gamma = \frac{m + bx_0}{(1 + m^2 - b^2)^{1/2}(1 - x_0^2)^{1/2}},$$

and the conclusion of (b) is established. The details are left as an exercise.

Example A1. Let $P = (1/2, 0), Q = (9/10, 0), Q = (3/4, 1/4)$. So $\overleftrightarrow{PQ}$ is the line $y = 0$ ($m = 0 = b$) and $\overrightarrow{PR}$ has the equation $y = x - \frac{1}{2}$ ($n = 1, c = -1/2$). (See Figure A1.)

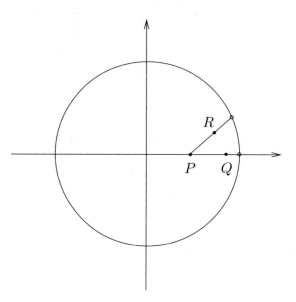

Figure A1

The Euclidean angle measure $\angle QPR = 45°$, but according to Proposition A1,

$$\text{(H-measure) } \angle QPR = \cos^{-1}\left(\frac{1 + 0 - 0}{\sqrt{1}\sqrt{1 + 1^2 - 1/4}}\right)$$

$$= \cos^{-1}\left(\frac{2}{\sqrt{7}}\right) = 40.89°.$$

This illustrates the "distortion" induced by the procedure for computing hyperbolic angle measure. Problem A2 shows that the difference between the Euclidean and hyperbolic measures of an angle is even greater if the vertex of the angle is closer to the boundary of H. In contrast, according to the next result, Euclidean and hyperbolic angle measures are equal when the angle's vertex is the origin.

Corollary A2. Let $O = (0,0), Q = (x_1, y_1), R = (x_2, y_2)$ be non-collinear points in $\mathbb{H}$. Then

$$(\mathbb{H}\text{-measure}) \angle QOR = (\mathbb{E}\text{-measure}) \angle QOR.$$

Proof. We give two proofs. The first is an informal observation of the sort that underlie our claims in Chapter 11 that $\mathbb{H}$ satisfies the axioms introduced in that section. The second, and perhaps more rigorous, argument is based on Proposition A1.

The projection of O onto the upper hemisphere is the point $(0, 0, 1)$. The two vectors tangent to the circular arcs above $\overrightarrow{OQ}$ and $\overrightarrow{OR}$, respectively, with their tails at $(0, 0, 1)$, both lie in the plane tangent to the hemisphere at $(0, 0, 1)$. This plane is horizontal; that is, it is parallel to the xy-plane, which contains the points of $\mathbb{H}$. Since $\overrightarrow{OQ}$ and $\overrightarrow{OR}$ project vertically onto these tangent vectors, the angles in the two planes are congruent. The conclusion follows.

We begin the second proof by noting that we may assume that $\overleftrightarrow{PQ}$ has the equation $y = mx$, and $\overleftrightarrow{PR}$ has equation either $y = nx$ or $x = 0$. Suppose that $x_1 > 0$ and that $x_2 > 0$ (if $\overleftrightarrow{PR}$ is nonvertical) or that $y_2 > 0$ (if $\overleftrightarrow{PR}$ is the y-axis). Then Proposition A1 implies that

$$\begin{aligned}
(\mathbb{H}\text{-measure}) \angle QOR \quad &= \quad \cos^{-1}\left(\frac{1 + mn}{\sqrt{1 + m^2}\sqrt{1 + n^2}}\right) \\
\text{or} \quad &\cos^{-1}\left(\frac{m}{\sqrt{1 + m^2}}\right).
\end{aligned}$$

On the other hand, the Law of Cosines applied to $\triangle QOR$ with $Q = (x_1, mx_1)$ and $R = (x_2, nx_2)$ or $(0, y_2)$ yields that

$$\begin{aligned}
(\mathbb{E}\text{-measure}) \angle QOR \quad &= \quad \cos^{-1}\left(\frac{1 + mn}{\sqrt{1 + m^2}\sqrt{1 + n^2}}\right) \\
\text{or} \quad &\cos^{-1}\left(\frac{m}{\sqrt{1 + m^2}}\right).
\end{aligned}$$

(The details are left as an exercise.) So in this case, $(\mathbb{E}\text{-measure})$ $\angle QOR = (\mathbb{H}\text{-measure}) \angle QOR$.

Suppose that $x_1 < 0$ or that $x_2 < 0$ or that $\overleftrightarrow{PR}$ is the y-axis and $y_2 < 0$. Then by Theorem 11.8, $\angle QOR$ is either supplementary or vertical (as both a hyperbolic and Euclidean angle) to an angle, as in the preceding paragraph. Hence, the conclusion holds in all cases.

Now we begin to establish the SAS property for $\mathbb{H}$. We adapt and make rigorous the principle of superposition, which Euclid invoked to justify that SAS holds in $\mathbb{E}$. This principle says that if $\triangle ABC$ and $\triangle XYZ$ are triangles with $\angle A = \angle X$, $AB = XY$ and $AC = XZ$, then we can "move" $\triangle ABC$ without changing the size of its sides or angles, and place it on top of $\triangle XYZ$ with A on X, B on Y, and C on Z; hence, $\triangle ABC \cong \triangle XYZ$. Nothing in either Euclid's axioms or our first 20 axioms renders such an operation meaningful, let alone permissible. Nevertheless, superposition is plausible enough for $\mathbb{E}$ (and for $\mathbb{S}$) that we simply assumed Axiom SAS for those models. But there is nothing obvious about how triangles can be "moved" in $\mathbb{H}$ without changing distances or angle measures. We introduce the concept of *symmetries* of $\mathbb{H}$ in order to clarify and develop this notion.

Definition. A *symmetry* of $\mathbb{H}$ is a function $\sigma : \mathbb{H} \to \mathbb{H}$ that preserves distance and angle measure. That is, for any distinct points P, Q, R in $\mathbb{H}$,

$$d_{\mathbb{H}}(\sigma(P)\sigma(Q)) = d_{\mathbb{H}}(PQ); \text{ and}$$

$$(\mathbb{H}\text{-measure}) \ \angle \sigma(Q)\sigma(P)\sigma(R) = (\mathbb{H}\text{-measure}) \ \angle QPR.$$

Note that if σ preserves distance, then for any points $P \neq Q$, $0 < d_{\mathbb{H}}(PQ) = d_{\mathbb{H}}(\sigma(P)\sigma(Q))$; hence, $\sigma(P) \neq \sigma(Q)$. So if P, Q, R are distinct, then $\sigma(P), \sigma(Q), \sigma(R)$ are also distinct and $\angle \sigma(P)\sigma(Q)\sigma(R)$ makes sense.

Next we find some symmetries of $\mathbb{H}$. We begin with a class of functions that preserve distance and angle measure in $\mathbb{E}$.

Example A2. Let l be a line in $\mathbb{E}$. The *reflection across l* is the function ϕ $(= \phi_l$; it depends on l) from $\mathbb{E}$ into $\mathbb{E}$ such that

$$\phi(P) = P \text{ for all points } P \in l;$$

and if $P \notin l$, let Q be the point on l such that $\overleftrightarrow{PQ} \perp l$ (Theorem 14.4). Let P' be the unique point on the opposite ray to $\overrightarrow{QP}$ such that $QP' = QP$. Then define $\phi(P) = P'$.

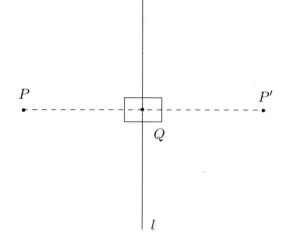

Figure A2

So ϕ sends each point to its "mirror image" across l. (See Figure A2.) It is easily shown from the results of Chapters 13–16 (primarily from Axiom SAS) that ϕ preserves Euclidean distance and angle measure and (consequently) sends any line onto another line. We omit the proof.

Proposition A3. Let l be any line in $\mathbb{E}$ that contains the origin O. Let $\phi = \phi_l$. Then $\phi(\mathbb{H}) = \mathbb{H}$ and ϕ is a symmetry of $\mathbb{H}$.

Proof. Let $\mathcal{C}$ denote the unit circle in $\mathbb{E}$. Thus, $\mathcal{C} = \{P \in \mathbb{E} : e(PO) = 1\}$ and $\mathbb{H} = \{P \in \mathbb{E} : e(PO) < 1\}$. Now $\phi(O) = O$ as $O \in l$. Since ϕ preserves distance in $\mathbb{E}$, it follows that $\phi(\mathcal{C}) = \mathcal{C}$ and $\phi(\mathbb{H}) = \mathbb{H}$.

Let P, Q be any two distinct points in $\mathbb{H}$; let M, N be the points where Euclidean line $\overleftrightarrow{PQ}$ meets $\mathcal{C}$, and with M-P-Q-N in $\mathbb{E}$. Then $\phi(P)$, $\phi(Q) \in \mathbb{H}$; $\phi(M), \phi(N) \in \mathcal{C}$; and since ϕ preserves collinearity and distance, $\phi(M)$-$\phi(P)$-$\phi(Q)$-$\phi(N)$. (See Figure A3.)

Since $e(\phi(P)\phi(N)) = e(PN), e(\phi(Q)\phi(M)) = e(QM)$, etc.,

$$
d_{\mathbb{H}}(\phi(P)\phi(Q)) = \ln\left(\frac{e(\phi(P)\phi(N))e(\phi(Q)\phi(M))}{e(\phi(P)\phi(M))e(\phi(Q)\phi(N))}\right)
$$

$$
= \ln\left(\frac{e(PN)e(QM)}{e(PM)e(QN)}\right) = d_{\mathbb{H}}(PQ),
$$

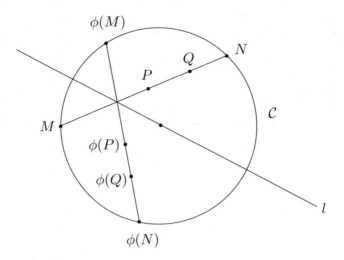

Figure A3

so that ϕ preserves $\mathbb{H}$-distance. To see that ϕ preserves $\mathbb{H}$-measure of angles, we make the following observations: The reflection of $\mathbb{E}$ across the line l is part of the reflection of $\mathbb{R}^3$ across the vertical plane that contains l. This reflection sends the upper unit hemisphere onto itself and preserves the angle between the tangent vectors to any pair of intersecting circular arcs on the hemisphere. Therefore, ϕ preserves $\mathbb{H}$-measure of angles.

Proposition A4. Suppose that $O = (0,0)$ and P, Q are distinct points of $\mathbb{H}$ such that $e(OP) = e(OQ)$. Let h be the ray with endpoint O that bisects $\angle POQ$ and let l be the carrier of h. (If $\angle POQ$ is a straight angle, let h be either ray with endpoint O on the line l with $l \perp \overleftrightarrow{PQ}$ at O.) Then $\phi_l(P) = Q$ and $d_{\mathbb{H}}(OP) = d_{\mathbb{H}}(OQ)$.

Proof. Let $h = \overrightarrow{OX}$. Let $\gamma = \angle POX = \angle QOX$. (See Figure A4.) Then

$$\angle \phi_l(P)OX = \angle \phi_l(P)\phi_l(O)\phi_l(X) = \angle POX = \angle QOX.$$

It follows from Corollary 12.3 that $\overrightarrow{O\phi_l(P)} = \overrightarrow{OQ}$. Since $e(O\phi_l(P)) = e(\phi_l(O)\phi_l(P)) = e(OP) = e(OQ)$, we have that $\phi_l(P) = Q$. Now by Proposition A3, $d_{\mathbb{H}}(OP) = d_{\mathbb{H}}(\phi_l(O)\phi_l(P)) = d_{\mathbb{H}}(OQ)$.

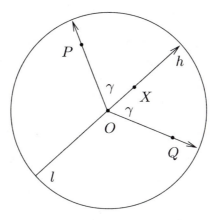

Figure A4

Definition. If σ, τ are any symmetries of $\mathbb{H}$, the *product* $\sigma\tau$ is defined as the composite function $\sigma \circ \tau$. That is,

$$\sigma\tau(P) = \sigma(\tau(P)), \text{ for all } P \in \mathbb{H}.$$

Proposition A5. If σ and τ are symmetries of $\mathbb{H}$, then so is $\sigma\tau$.

Proof. Since $\sigma(\mathbb{H}) \subseteq \mathbb{H}$ and $\tau(\mathbb{H}) \subseteq \mathbb{H}$, it follows that $\sigma\tau(\mathbb{H}) \subseteq \mathbb{H}$. For any points $P, Q \in \mathbb{H}$,

$$d_{\mathbb{H}}(\sigma\tau(P)\sigma\tau(Q)) = d_{\mathbb{H}}(\sigma(\tau(P))\sigma(\tau(Q))) \text{ (by definition of } \sigma\tau)$$
$$= d_{\mathbb{H}}(\tau(P)\tau(Q)) = d_{\mathbb{H}}(PQ) \text{ (since } \sigma \text{ and } \tau \text{ preserve } \mathbb{H}\text{-distance).}$$

Therefore, $\sigma\tau$ preserves $\mathbb{H}$-distance. The proof that $\sigma\tau$ preserves $\mathbb{H}$-measure of angles is similar.

Proposition A6. If σ is a symmetry of $\mathbb{H}$ and $\triangle ABC$ is in $\mathbb{H}$, then $\sigma(A), \sigma(B), \sigma(C)$ are noncollinear points in $\mathbb{H}$ and $\triangle\sigma(A)\sigma(B)\sigma(C) \cong \triangle ABC$.

The proof is immediate.

Example A3. Let l, m be Euclidean lines through $O = (0,0)$. Suppose that l is the carrier of $\overrightarrow{OP}$, m is the carrier of $\overrightarrow{OQ}$, and $\angle POQ = \theta \leq 90$. Then, as can be seen from Figure A5, $\phi_m\phi_l$ is a rotation about O through an angle of measure 2θ.

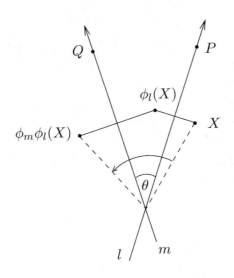

Figure A5

Definition. The *identity function* $id : \mathbb{H} \to \mathbb{H}$ that fixes every point of $\mathbb{H}$ (that is, $id(P) = P$ for all $P \in \mathbb{H}$) is clearly a symmetry of $\mathbb{H}$. Note that $id = \phi_l\phi_l$ for any line l through O.

We have found two types of symmetries of $\mathbb{H}$, reflections across a line through O and rotations about O. We shall need more symmetries in order to implement the following strategy:

Given $\triangle ABC$ and $\triangle XYZ$ in $\mathbb{H}$ with ($\mathbb{H}$-measure) $\angle A = \angle X$, $d_{\mathbb{H}}(AB) = d_{\mathbb{H}}(XY)$ and $d_{\mathbb{H}}(AC) = d_{\mathbb{H}}(XZ)$, then we find symmetries σ, τ of $\mathbb{H}$ with $\sigma(A) = \tau(X)$, $\sigma(B) = \tau(Y)$, $\sigma(C) = \tau(Z)$. Proposition A6 will imply that

$$\triangle ABC \cong \triangle\sigma(A)\sigma(B)\sigma(C) \cong \triangle\tau(X)\tau(Y)\tau(Z) \cong \triangle XYZ.$$

We introduce the following functions and prove the not at all obvious claim that they are symmetries of $\mathbb{H}$. They are *not* symmetries of $\mathbb{E}$.

Definition. Fix any positive real number λ with $\lambda \neq 1$. Define the function μ $(= \mu_\lambda;$ it depends on $\lambda)$ from (a subset of) $\mathbb{E}$ into $\mathbb{E}$ by

$$\mu(x, y) = \left(\frac{2\lambda x}{(1 - y) + \lambda^2(1 + y)}, \frac{\lambda^2(1 + y) - (1 - y)}{(1 - y) + \lambda^2(1 + y)} \right).$$

The formula that defines the function μ_λ does not at first glance seem particularly natural or illuminating. But it comes from a sequence of rather simple geometric procedures that we shall describe after we prove some facts about μ.

Calculations for μ are simplified by the following notation:

Definition. Fix $\lambda > 0$ with $\lambda \neq 1$. For any $y \in \mathbb{R}$, define

$$u_y = (1 - y) + \lambda^2(1 + y) = (\lambda^2 + 1) + y(\lambda^2 - 1).$$

Then the formula for μ can be written as

$$\mu(x, y) = \left(\frac{2\lambda x}{u_y}, \frac{u_y - 2(1 - y)}{u_y} \right).$$

In Propositions A7–11 and A13–14, λ remains a fixed real number with $\lambda > 0$, $\lambda \neq 1$.

Proposition A7. For any $y, z \in \mathbb{R}$,

(a) $u_y - u_z = (y - z)(\lambda^2 - 1)$;   (b) $u_y(1 - z) - u_z(1 - y) = 2\lambda^2(y - z)$.

Furthermore, if $w, m, b, p, q, r \in \mathbb{R}$ with $y = mp + b$, $z = mq + b$, $w = mr + b$, then

(c) $u_y = pm(\lambda^2 - 1) + u_b$;   (d) $(q - p)u_w + (r - q)u_y = (r - p)u_z$.

Proof. (a), (b), (c) follow easily from the definitions of u_y, u_z, u_b and are left as an exercise. To establish (d), note that by (c),

$$(q - p)u_w + (r - q)u_y$$
$$= (q - p)(rm(\lambda^2 - 1) + u_b) + (r - q)(pm(\lambda^2 - 1) + u_b)$$
$$= (r - p)(qm(\lambda^2 - 1) + u_b) = (r - p)u_z.$$

Proposition A8. (a) The domain of μ is all of $\mathbb{E}$ except for the horizontal line $y = \dfrac{1 + \lambda^2}{1 - \lambda^2}$.

(b) $\mu(\mathbb{H}) \subseteq \mathbb{H}$ and $\mu(\mathcal{C}) \subseteq \mathcal{C}$, where $\mathcal{C}$ denotes the unit circle.

(c) For any Euclidean line l, let $l^* = l \cap$ domain μ. Then $\mu(l^*)$ is contained in a Euclidean line. Specifically,

If l has the equation $y = mx + b$ and $b \neq \dfrac{1+\lambda^2}{1-\lambda^2}$, then all points (x, y) in $\mu(l^*)$ satisfy the equation $y = Mx + B$, where $M = \dfrac{2m\lambda}{u_b}$, $B = \dfrac{u_b + 2b - 2}{u_b}$.

If l has the equation $y = mx + b$ with $m \neq 0$ and $b = \dfrac{1+\lambda^2}{1-\lambda^2}$, then $\mu(l^*)$ is contained in the vertical line $x = \dfrac{2\lambda}{(\lambda^2-1)m}$.

If l is the vertical line $x = a \neq 0$, then $\mu(l^*)$ is contained in the line $y = Mx + B$, where $M = \dfrac{-2\lambda}{(\lambda^2-1)a}$ and $B = \dfrac{\lambda^2+1}{\lambda^2-1}$.

If l is the y-axis $(x = 0)$, then $\mu(l^*)$ is contained in l.

Proof. (a) $\mu(x, y)$ is defined $\Leftrightarrow u_y \neq 0 \Leftrightarrow y \neq \dfrac{1+\lambda^2}{1-\lambda^2}$.

(b) For any $(x, y) \in$ domain μ, let s denote the square of the Euclidean length of $\mu(x, y)$. Thus

$$s = \left(\frac{2\lambda x}{u_y}\right)^2 + \left(\frac{u_y - 2(1-y)}{u_y}\right)^2$$

$$= \frac{4\lambda^2 x^2 + u_y^2 - 4u_y(1-y) + 4(1-y)^2}{u_y^2}$$

$$= \frac{4\lambda^2 x^2 + u_y^2 - 4(1-y)^2 - 4\lambda^2(1-y^2) + 4(1-y)^2}{u_y^2}$$

$$= \frac{4\lambda^2(x^2+y^2) + u_y^2 - 4\lambda^2}{u_y^2}.$$

Now $(x, y) \in \mathbb{H} \Leftrightarrow x^2 + y^2 < 1 \Leftrightarrow s < \dfrac{4\lambda^2 + u_y^2 - 4\lambda^2}{u_y^2} = 1$; and

$(x, y) \in \mathcal{C} \Leftrightarrow x^2 + y^2 = 1 \Leftrightarrow s = \dfrac{4\lambda^2 + u_y^2 - 4\lambda^2}{u_y^2} = 1$. The claims of (b) follow.

(c) Suppose that l has the equation $y = mx + b$, where $b \neq \dfrac{1 + \lambda^2}{1 - \lambda^2}$.

Since $\mu(x, y) = \left(\dfrac{2\lambda x}{u_y}, \dfrac{u_y - 2(1 - y)}{u_y} \right)$, we seek constants M, B such that

$$\frac{u_y - 2(1 - y)}{u_y} = M \left(\frac{2\lambda x}{u_y} \right) + B,$$

(*)

for all x except where $y = mx + b = \dfrac{1 + \lambda^2}{1 - \lambda^2}$.

Note that by Proposition A7(c), (*) holds $\Leftrightarrow$

$$u_y - 2(1 - y) = M2\lambda x + Bu_y$$

$$\Leftrightarrow \quad xm(\lambda^2 - 1) + u_b - 2(1 - mx - b)$$
$$= M2\lambda x + Bxm(\lambda^2 - 1) + Bu_b$$

$$\Leftrightarrow \quad m(\lambda^2 + 1)x + u_b + 2b - 2 = (2\lambda M + Bm(\lambda^2 - 1))x + Bu_b$$

$$\Leftrightarrow \quad B = \frac{u_b + 2b - 2}{u_b} \quad \text{and}$$

$$M = \frac{m(\lambda^2 + 1 + B(1 - \lambda^2))}{2\lambda}$$

$$= \frac{m}{2\lambda u_b}((\lambda^2 + 1)u_b + (u_b + 2b - 2)(1 - \lambda^2))$$

$$= \frac{m}{2\lambda u_b} \cdot 4\lambda^2 = \frac{2m\lambda}{u_b}.$$

This proves the first case in part (c). The proof of the remaining three cases is similar and is left as an exercise.

Now we give the promised geometric interpretation of the function μ. Consider a point $(x, y) \in \mathbb{H}$. Let Q denote the projection of (x, y) vertically onto the upper unit hemisphere $\mathcal{U}$, so $Q = (x, y, \sqrt{1 - x^2 - y^2})$. Let $\mathcal{K}$ denote the vertical halfplane in $\mathbb{R}^3$ defined by $y = 1$, $z > 0$. Let $P = (0, -1, 0)$. The *stereographic projection* of $\mathcal{U}$ onto $\mathcal{K}$, using the point P, goes as follows:

The line through P and Q in $\mathbb{R}^3$ is parametrized as

$$(xt, (y + 1)t - 1, t\sqrt{1 - x^2 - y^2}) \text{ for } t \in \mathbb{R}.$$

This line meets $\mathcal{K}$ at the point R where $(y+1)t-1 = 1$, so $t = \dfrac{2}{y+1}$.

Thus $R = \left(\dfrac{2x}{y+1}, 1, \dfrac{2\sqrt{1-x^2-y^2}}{y+1}\right)$. Note that $\dfrac{2\sqrt{1-x^2-y^2}}{y+1} > 0$, as $y > -1$. The correspondence $Q \mapsto R$ is called the stereographic projection of $\mathcal{U}$ onto $\mathcal{K}$. (See Figure A6.)

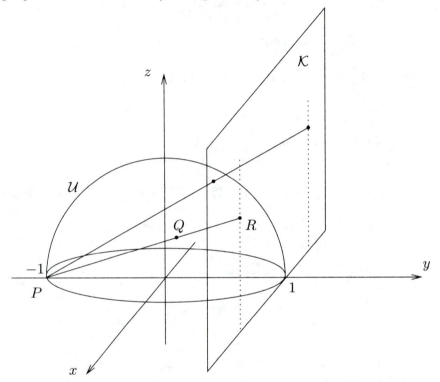

Figure A6

To go from $\mathcal{K}$ back to $\mathcal{U}$, start with $R = (x, 1, z) \in \mathcal{K}$. The line through R and $P = (0, -1, 0)$ is parametrized as $(xt, 2t-1, zt)$ for $t \in \mathbb{R}$. The line meets $\mathcal{U}$ where

$$(xt)^2 + (2t-1)^2 + (zt)^2 = 1 \Leftrightarrow x^2t^2 + 4t^2 - 4t + z^2t^2 = 0$$
$$\Leftrightarrow t = \frac{4}{x^2 + z^2 + 4}.$$

Hence, the line meets $\mathcal{U}$ at the point

$$Q = \left(\frac{4x}{x^2+z^2+4}, \frac{4-x^2-z^2}{x^2+z^2+4}, \frac{4z}{x^2+z^2+4}\right).$$

Define the function γ with domain $\mathbb{H}$ such that

$$\gamma(x, y) = \left(\frac{2x}{y+1}, 1, \frac{2\sqrt{1 - x^2 - y^2}}{y+1} \right), \quad \text{all } (x, y) \in \mathbb{H};$$

and define the function β with domain $\mathcal{K}$ such that

$$\beta(x, 1, z) = \left(\frac{4x}{x^2 + z^2 + 4}, \frac{4 - x^2 - z^2}{x^2 + z^2 + 4} \right) \quad \text{all } (x, 1, z) \in \mathcal{K}.$$

It is immediate from the preceding observations, or it can be checked directly by algebra, that

$$\beta \circ \gamma(x, y) = (x, y), \quad \text{for all } (x, y) \in \mathbb{H}.$$

Recall that we have fixed the real number $\lambda > 0$, with $\lambda \neq 1$. Let $\sigma = \sigma_\lambda$ be the function from $\mathcal{K}$ onto $\mathcal{K}$ such that $\sigma(x, 1, z) = (\lambda x, 1, \lambda z)$ for all $(x, 1, z) \in \mathcal{K}$. Thus, σ "expands" (or "contracts") $\mathcal{K}$ by the scaling factor λ.

The composite function $\beta \circ \sigma \circ \gamma$ takes the point $(x, y) \in \mathbb{H}$ and sends it first to the stereographic projection of $(x, y, \sqrt{1 - x^2 - y^2})$ in $\mathcal{K}$, then multiplies the first and third coordinates of this point by the scaling factor λ, and then takes the reverse projection of *this* point to $\mathcal{U}$ and back down to $\mathbb{H}$. It is a straightforward exercise to verify the following:

Proposition A9. $\beta \circ \sigma \circ \gamma(x, y) = \mu(x, y)$, for all $(x, y) \in \mathbb{H}$.

The correspondence γ between $\mathbb{H}$ and $\mathcal{K}$ may be used to make the halfplane $\mathcal{K}$ into another example of a hyperbolic plane, one that is equivalent to $\mathbb{H}$. This halfplane model is treated in a number of other texts.

Proposition A10. Suppose that $P = (p, y)$ and $Q = (q, z)$ are in the domain of μ.

(a) If $\overleftrightarrow{PQ}$ has the equation $y = mx + b$ and $b \neq \dfrac{1 + \lambda^2}{1 - \lambda^2}$, let M be the slope of $\overleftrightarrow{\mu(P)\mu(Q)}$ as in Proposition A8. Then

$$e(\mu(P)\mu(Q)) = \frac{2\lambda |p - q| |u_b| \sqrt{1 + M^2}}{|u_y u_z|}.$$

(b) If $\overleftrightarrow{PQ}$ has the equation $y = mx + b$ and $b = \dfrac{1 + \lambda^2}{1 - \lambda^2}$, or if $\overleftrightarrow{PQ}$ is the y-axis, then

$$e(\mu(P)\mu(Q)) = \frac{4\lambda^2 |y - z|}{|u_y u_z|}.$$

(c) If $\overleftrightarrow{PQ}$ is the vertical line $x = a \neq 0$, let M be the slope of $\overleftrightarrow{\mu(P)\mu(Q)}$ as in Proposition A8. Then

$$e(\mu(P)\mu(Q)) = \frac{2\lambda |a(\lambda^2 - 1)||y - z|\sqrt{1 + M^2}}{|u_y u_z|}.$$

Proof. (a) Proposition A8(c) implies that $\overleftrightarrow{\mu(P)\mu(Q)}$ is a line with slope $M = 2m\lambda/u_b$. The first coordinates of $\mu(P), \mu(Q)$ are $2\lambda p/u_y$ and $2\lambda q/u_z$, respectively. Propositions 1.1 and A7(c) yield that

$$e(\mu(P)\mu(Q)) = \left| \frac{2\lambda p}{u_y} - \frac{2\lambda q}{u_z} \right| \sqrt{1 + M^2} = \frac{2\lambda |p u_z - q u_y| \sqrt{1 + M^2}}{|u_y u_z|}$$
$$= 2\lambda |p(qm(\lambda^2 - 1) + u_b) - q(pm(\lambda^2 - 1) + u_b)| \sqrt{1 + M^2}/|u_y u_z|$$
$$= 2\lambda |p - q||u_b| \sqrt{1 + M^2}/|u_y u_z|.$$

(b) The hypotheses of (b) and Proposition A8(c) imply that $\overleftrightarrow{\mu(P)\mu(Q)}$ is a vertical line. The second coordinates of $\mu(P), \mu(Q)$ are $(u_y - 2(1 - y))/u_y$ and $(u_z - 2(1 - z))/u_z$, respectively. Hence, by Proposition A7(b),

$$e(\mu(P)\mu(Q)) = \left| 1 - \frac{2(1 - y)}{u_y} - \left(1 - \frac{2(1 - z)}{u_z} \right) \right|$$
$$= 2 \left| \frac{1 - z}{u_z} - \frac{1 - y}{u_y} \right| = 2|u_y(1 - z) - u_z(1 - y)|/|u_y u_z|$$
$$= 4\lambda^2 |y - z|/|u_y u_z|.$$

The proof of (c) is left as an exercise.

Proposition A11. Suppose that $P = (p, y)$, $Q = (q, z)$, $R = (r, w)$ are distinct, collinear points in $\mathbb{E}$ with P-Q-R. Suppose also that if

$\lambda < 1$, then each of y, z, w is less than $(1 + \lambda^2)/(1 - \lambda^2)$; and if $\lambda > 1$, then each of y, z, w is greater than $(1 + \lambda^2)/(1 - \lambda^2)$. Then $\mu(P)$-$\mu(Q)$-$\mu(R)$.

Proof. Let s denote any of y, z, w. If $\lambda < 1$, then $s < (1+\lambda^2)/(1-\lambda^2)$ by hypothesis. Hence $u_s = (\lambda^2 + 1) + s(\lambda^2 - 1) > 0$. If $\lambda > 1$, then $s > (1+\lambda^2)/(1-\lambda^2)$ and $u_s < 0$. So in either case u_y, u_z, u_w all have the same sign. Thus each of $u_y u_z, u_z u_w, u_y u_w$ is positive.

We treat the case where $\overleftrightarrow{PQ}$ is a line with equation $y = mx + b$ and $b \neq (1 + \lambda^2)/(1 - \lambda^2)$. The proof for the other cases of $\overleftrightarrow{PQ}$ is similar and is left as an exercise.

Since P-Q-R implies R-Q-P, we may assume that $p < q < r$. By Proposition A8(c), $\mu(P), \mu(Q), \mu(R)$ are all on one line with slope $M = 2m\lambda/u_b$. By Propositions A10(a) and A7(d), $\mu(P), \mu(Q), \mu(R)$ are distinct and

$$e(\mu(P)\mu(Q)) + e(\mu(Q)\mu(R)) = 2\lambda|u_b|\sqrt{1 + M^2}\left(\frac{|p - q|}{|u_y u_z|} + \frac{|q - r|}{|u_z u_w|}\right)$$

$$= 2\lambda|u_b|\sqrt{1 + M^2}\left(\frac{(q - p)}{u_y u_z} + \frac{(r - q)}{u_z u_w}\right)$$

$$= 2\lambda|u_b|\sqrt{1 + M^2}\left(\frac{(q - p)u_w + (r - q)u_y}{u_y u_z u_w}\right)$$

$$= 2\lambda|u_b|\sqrt{1 + M^2}\left(\frac{(r - p)u_z}{u_y u_z u_w}\right)$$

$$= 2\lambda|u_b|\sqrt{1 + M^2}|p - r|/|u_y u_w| = e(\mu(P)\mu(R)).$$

Therefore, $\mu(P)$-$\mu(Q)$-$\mu(R)$.

Theorem A12. Fix any positive real number λ with $\lambda \neq 1$, and let $\mu = \mu_\lambda$ as defined previously. If P, Q are any two points in $\mathbb{H}$, then

$$d_{\mathbb{H}}(PQ) = d_{\mathbb{H}}(\mu(P)\mu(Q)),$$

so that μ preserves distance in $\mathbb{H}$.

Proof. Let $P = (p, y)$, $Q = (q, z)$. Let $S = (s, t)$ and $R = (r, w)$ be the points where $\overleftrightarrow{PQ}$ meets the unit circle $\mathcal{C}$, and so that S-P-Q-R.

Since the second coordinates of all four points are between -1 and 1, the hypotheses of Proposition A11 are satisfied by any three of these points. So $\mu(S)$-$\mu(P)$-$\mu(Q)$-$\mu(R)$ by Proposition A11. Since $\mu(S)$ and $\mu(R)$ are in $\mathcal{C}$, the definition of hyperbolic distance yields that

$$d_{\mathbb{H}}(\mu(P)\mu(Q)) = \ln\left(\frac{e(\mu(P)\mu(R))e(\mu(Q)\mu(S))}{e(\mu(P)\mu(S))e(\mu(Q)\mu(R))}\right).$$

Suppose that $\overleftrightarrow{PQ}$ is a nonvertical line whose y-intercept is $b \neq (1+\lambda^2)/(1-\lambda^2)$. Then by Proposition A10(a), where M is the slope of $\overleftrightarrow{\mu(P)\mu(Q)}$, and by Proposition 1.1,

$$\frac{e(\mu(P)\mu(R))e(\mu(Q)\mu(S))}{e(\mu(P)\mu(S))e(\mu(Q)\mu(R))}$$

$$= \frac{\dfrac{2\lambda|p-r||u_b|\sqrt{1+M^2}}{|u_y u_w|} \cdot \dfrac{2\lambda|q-s||u_b|\sqrt{1+M^2}}{|u_z u_t|}}{\dfrac{2\lambda|p-s||u_b|\sqrt{1+M^2}}{|u_y u_t|} \cdot \dfrac{2\lambda|q-r||u_b|\sqrt{1+M^2}}{|u_z u_w|}}$$

$$= \frac{|p-r||q-s|}{|p-s||q-r|} = \frac{e(PR)e(QS)}{e(PS)e(QR)},$$

so it follows that $d_{\mathbb{H}}(\mu(P)\mu(Q)) = d_{\mathbb{H}}(PQ)$. The proof for the other cases of $\overleftrightarrow{PQ}$ follows in a similar way from Proposition A10(b, c).

The next result is a preliminary step for Proposition A14, which says that μ in fact preserves hyperbolic measure of angles.

Proposition A13. For any angle $\angle PQR$ in $\mathbb{H}$, with ($\mathbb{H}$-measure) $\angle PQR$, then

$$\cos(\angle\mu(P)\mu(Q)\mu(R)) = \pm\cos(\angle PQR),$$

so that $\angle\mu(P)\mu(Q)\mu(R) = \angle PQR$ or $180 - \angle PQR$.

Proof. Since μ preserves collinearity and betweenness relations for points, μ preserves zero angles and straight angles. Hence, μ sends two supplementary angles with the same vertex to another pair of supplementary angles; and μ sends any pair of vertical angles to another pair of vertical angles. So it suffices to assume that $\angle PQR$

satisfies $P = (p, y), Q = (q, z), R = (r, w)$ with $p > q$ and either $r > q$, or $r = q$ with $w > z$. (See Figure A7.)

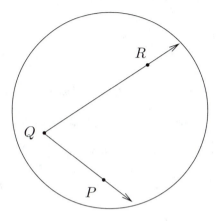

Figure A7

We consider here the case where $p > q$ and $r > q$. We also assume that $\overleftrightarrow{QP}$ has the equation $y = mx + b$ and $\overleftrightarrow{QR}$ has the equation $y = nx + c$ with $b \neq (1 + \lambda^2)/(1 - \lambda^2)$ and $c \neq (1 + \lambda^2)/(1 - \lambda^2)$. The proof for the other cases is similar and is omitted.

By Proposition A8, $\overleftrightarrow{\mu(Q)\mu(P)}$ and $\overleftrightarrow{\mu(Q)\mu(R)}$ has the equations $y = Mx + B$ and $y = Nx + C$, respectively, where $M = 2m\lambda/u_b$, $B = (u_b + 2b - 2)/u_b$, $N = 2n\lambda/u_c$, $C = (u_c + 2c - 2)/u_c$. Hence by Proposition A1, the cosine of ($\mathbb{H}$-measure) $\angle \mu(P)\mu(Q)\mu(R)$ is

$$(\pm) \frac{1 + MN - BC}{\sqrt{1 + M^2 - B^2}\sqrt{1 + N^2 - C^2}} .$$

Now

$$
\begin{aligned}
\sqrt{1 + M^2 - B^2} &= \left(u_b^2 + 4m^2\lambda^2 - (u_b + 2b - 2)^2 \right)^{1/2} \Big/ |u_b| \\
&= \left(4m^2\lambda^2 - 4(b - 1)(u_b + b - 1) \right)^{1/2} \Big/ |u_b| \\
&= \left(4m^2\lambda^2 - 4(b - 1)(b + 1)\lambda^2 \right)^{1/2} \Big/ |u_b| \\
&= 2\lambda \left(1 + m^2 - b^2 \right)^{1/2} \Big/ |u_b| ;
\end{aligned}
$$

and similarly,

$$\sqrt{1 + N^2 - C^2} = 2\lambda \left(1 + n^2 - c^2\right)^{1/2} / \, |u_c| \, .$$

Also, the definitions of u_b and u_c yield

$$
\begin{aligned}
1 &+ MN - BC \\
&= \left(u_b u_c + 4\lambda^2 mn - (u_b + 2b - 2)(u_c + 2c - 2)\right) / \, u_b u_c \\
&= \left(4\lambda^2 mn - 2(b-1)u_c - 2(c-1)u_b - 4(b-1)(c-1)\right) / \, u_b u_c \\
&= \left(4\lambda^2 mn + 4\lambda^2 - 4\lambda^2 bc\right) / \, u_b u_c \, .
\end{aligned}
$$

It follows that the cosine of ($\mathbb{H}$-measure) $\angle \mu(P)\mu(Q)\mu(R)$ equals

$$\pm \frac{1 + mn - bc}{(1 + m^2 - b^2)^{1/2}(1 + n^2 - c^2)^{1/2}} = \pm \cos(\angle PQR).$$

Proposition A14. For any angle $\angle PQR$ in $\mathbb{H}$,

$$(\mathbb{H}\text{-measure}) \ \angle \mu(P)\mu(Q)\mu(R) = \ (\mathbb{H}\text{-measure}) \ \angle PQR.$$

Proof. We know that μ preserves zero angles and straight angles. By Proposition A13, $\angle \mu(P)\mu(Q)\mu(R) = \angle PQR$ or $180 - \angle PQR$; hence μ preserves right angles. Since μ sends supplementary angles with a common vertex to another pair of supplementary angles, it suffices to assume that $\angle PQR$ is acute. Then by Theorem 11.6 applied to the fan $\overrightarrow{QP}\overrightarrow{QR}$, there is a ray $\overrightarrow{QT}$ with $\overrightarrow{QP}$-$\overrightarrow{QR}$-$\overrightarrow{QT}$ and $\angle PQT = 90$. Then $\angle \mu(P)\mu(Q)\mu(T) = 90$.

By the Crossbar Theorem (Theorem 12.4), there is an interior point X of $\overrightarrow{QR}$ with P-X-T. (See Figure A8.) Since μ preserves rays, $\overrightarrow{\mu(Q)\mu(X)} = \overrightarrow{\mu(Q)\mu(R)}$. Since $\mu(P)$-$\mu(X)$-$\mu(T)$ (Proposition A11), Axiom C yields that $\overrightarrow{\mu(Q)\mu(P)}$-$\overrightarrow{\mu(Q)\mu(X)}$-$\overrightarrow{\mu(Q)\mu(T)}$. Therefore,

$$\angle \mu(P)\mu(Q)\mu(R) = \angle \mu(P)\mu(Q)\mu(X) < \angle \mu(P)\mu(Q)\mu(T) = 90.$$

So $\angle PQR$ and $\angle \mu(P)\mu(Q)\mu(R)$ are both acute. Now Proposition A13 implies that $\angle \mu(P)\mu(Q)\mu(R) = \angle PQR$.

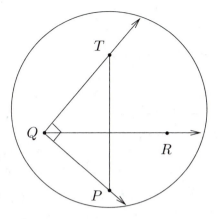

Figure A8

Combining Theorem A12 and Proposition A14, we have the following:

Theorem A15. For any $\lambda > 0$ with $\lambda \neq 1$, let $\mu = \mu_\lambda$ as defined above. Then μ is a symmetry of $\mathbb{H}$.

Now we have enough symmetries of $\mathbb{H}$ to implement the strategy outlined above and to reach the goal of this appendix.

Theorem A16. Suppose that $\triangle ABC$ and $\triangle XYZ$ are two triangles in $\mathbb{H}$ such that ($\mathbb{H}$-measure) $\angle A = \angle X$, $d_\mathbb{H}(AB) = d_\mathbb{H}(XY)$ and $d_\mathbb{H}(AC) = d_\mathbb{H}(XZ)$. Then there exist symmetries σ, τ of $\mathbb{H}$ such that $\sigma(A) = \tau(X)$, $\sigma(B) = \tau(Y)$, $\sigma(C) = \tau(Z)$. Hence, $\triangle ABC \cong \triangle XYZ$. So $\mathbb{H}$ satisfies the SAS property.

Proof. Let $O = (0,0)$. Fix points R, S in $\mathbb{H}$ so that R is on the positive y-axis and S is on the positive x-axis. We shall show first that there are points $P = (p,0)$ and $Q = (q,z)$ in $\mathbb{H}$ with $p > 0$ and $z > 0$, and so that there is a symmetry σ of $\mathbb{H}$ with $\sigma(A) = O$, $\sigma(B) = P$, and $\sigma(C) = Q$.

We begin by finding a symmetry σ_1 such that $\sigma_1(A) = O$. If $A = O$, we may let σ_1 be the identity function *id*. Otherwise, Proposition A4 yields that there is a line l through O such that $\phi_l(A)$ is on $\overrightarrow{OR}$ (l contains a ray that bisects $\angle AOR$). So $\phi_l(A) = (0, y)$ for some $0 < y < 1$.

Let $\lambda = (1-y)^{1/2}/(1+y)^{1/2}$, and $\mu = \mu_\lambda$. Then $\mu(0,y) = (0,0)$.

Define $\sigma_1 = \mu\phi_l$. Then

$$\sigma_1(A) = \mu(\phi_l(A)) = \mu(0, y) = O.$$

Let line t be the carrier of the bisecting ray for $\angle SO\sigma_1(B)$. Then $\phi_t(O) = O$ and $\phi_t(\sigma_1(B))$ is on $\overrightarrow{OS}$, again by Proposition A4. Let $P = \phi_t(\sigma_1(B))$. So $P = (p, 0)$ for some $0 < p < 1$. Now $\phi_t\sigma_1(C)$ is in $\mathbb{H}$ and is not collinear with $\phi_t\sigma_1(A) = O$ and $\phi_t\sigma_1(B) = P$. Hence, $\phi_t\sigma_1(C) = (q, z)$ for some q, z with $q^2 + z^2 < 1$ and $z \neq 0$.

If $z > 0$, define $Q = (q, z)$ and $\sigma = \phi_t\sigma_1$. Then $\sigma(A) = O$, $\sigma(B) = P$, $\sigma(C) = Q$ as desired.

If $z < 0$, define $Q = (q, -z)$, let ϕ denote the reflection across the x-axis, and let $\sigma = \phi\phi_t\sigma_1$. Then $\sigma(A) = \phi(\phi_t\sigma_1(A)) = \phi(O) = O$, $\sigma(B) = \phi(\phi_t\sigma_1(B)) = \phi(P) = P$, and $\sigma(C) = \phi(\phi_t\sigma_1(C)) = \phi(q, z) = (q, -z) = Q$ with $-z > 0$, also as desired. Note that σ is indeed a symmetry of $\mathbb{H}$ by Proposition A5.

The argument just completed also shows that there is a symmetry τ of $\mathbb{H}$ and points $P' = (p', 0)$, $Q' = (q', z')$ in $\mathbb{H}$ with $p' > 0$, $z' > 0$, and such that $\tau(X) = O$, $\tau(Y) = P'$, $\tau(Z) = Q'$. Now as $\overrightarrow{OP} = \overrightarrow{OP'} = \overrightarrow{OS}$, $\angle CAB = \angle ZXY$ by hypothesis, and σ, τ are symmetries of $\mathbb{H}$,

$$\angle Q'OS = \angle Q'OP' = \angle \tau(Z)\tau(X)\tau(Y) = \angle ZXY = \angle CAB$$
$$= \angle \sigma(C)\sigma(A)\sigma(B) = \angle QOP = \angle QOS.$$

Since $z > 0$ and $z' > 0$, $\overrightarrow{OQ}^o$ and $\overrightarrow{OQ'}^o$ lie in the same halfplane of $\mathbb{H}$ with edge $\overleftrightarrow{OS}$. As $\angle Q'OS = \angle QOS$, Corollary 12.3 implies that $\overrightarrow{OQ} = \overrightarrow{OQ'}$. Since

$$d_{\mathbb{H}}(OQ) = d_{\mathbb{H}}(\sigma(A)\sigma(C)) = d_{\mathbb{H}}(AC) = d_{\mathbb{H}}(XZ)$$
$$= d_{\mathbb{H}}(\tau(X)\tau(Z)) = d_{\mathbb{H}}(OQ'),$$

Theorem 8.6 yields that $Q = Q'$. Similarly, $P = P'$. Thus, $\sigma(A) = \tau(X)$, $\sigma(B) = \tau(Y)$, $\sigma(C) = \tau(Z)$. Since σ, τ are symmetries, it follows that

$$\triangle ABC \cong \triangle \sigma(A)\sigma(B)\sigma(C) \cong \triangle \tau(X)\tau(Y)\tau(Z) \cong \triangle XYZ.$$

Theorem A16 is proved.

Problem Set AII

1. Complete the details for the proof of Proposition A1.

2. Complete the details for the proof of Corollary A2.

3. Let $P = (.8, 0)$, $Q = (.9, 0)$, $R = (.9, .1)$. Compute both the $\mathbb{E}$-measure and the $\mathbb{H}$-measure of $\angle QPR$. Repeat for $P = (.98, 0)$, $Q = (.99, 0)$, $R = (.99, .01)$.

4. Suppose that A, B, X, Y are points in $\mathbb{H}$ such that $d_{\mathbb{H}}(AB) = d_{\mathbb{H}}(XY)$. Prove that there exist symmetries σ, τ of $\mathbb{H}$ such that $\sigma(A) = \tau(X)$ and $\sigma(B) = \tau(Y)$. (Hint: Study the proof of Theorem A16.)

5. Complete the proof of Proposition A5.

6. Let $O = (0, 0)$, $P = (p, 0)$ for some $0 < p < 1$ and $Q = (q, z)$ for some $z > 0$ with $q^2 + z^2 < 1$. Suppose that σ is a symmetry of $\mathbb{H}$ such that $\sigma(O) = O$, $\sigma(P) = P$, and with Q and $\sigma(Q)$ on the same side of $\overleftrightarrow{OP}$. Prove that $\sigma(Q) = Q$ and then that $\sigma(X) = X$ for all $X \in \mathbb{H}$.

7-13. Complete the proofs of Propositions/Theorems A7–A13.

14. Let l be the line in $\mathbb{H}$ whose Euclidean line is the x-axis (so $l = \{(x, 0) : -1 < x < 1\}$). Let q be any line in $\mathbb{H}$.

    (a) Prove that there exists a symmetry σ of $\mathbb{H}$ such that $\sigma(q) = l$. (Hint: See Problem 4.)

    (b) Prove that σ extends to a function such that $\sigma(\mathcal{C}) = \mathcal{C}$ (where $\mathcal{C}$ denotes the unit circle) and that σ sends the points where $\mathcal{C}$ meets the Euclidean line containing q to the points $(\pm 1, 0)$.

15. Let l be as in Problem 14 and let $(x_0, 0)$ be any point on l. Use Proposition A1 to show that the line in $\mathbb{H}$ that is perpendicular to l at $(x_0, 0)$ is the vertical line with Euclidean equation $x = x_0$.

16. Let l be as in Problem 14. Let s be any line in $\mathbb{H}$ such that $s \cap l = \emptyset$ and the Euclidean line containing s does not go through

$(1,0)$ or $(-1,0)$ (see Figure A9). Prove that there is a line in $\mathbb{H}$ that is perpendicular to both l and s. (Hint: Show first that s has a Euclidean equation $y = nx + c$ for some $n, c \in \mathbb{R}$ with $|n| < |c|$. Then use Problem 15 and Proposition A1.)

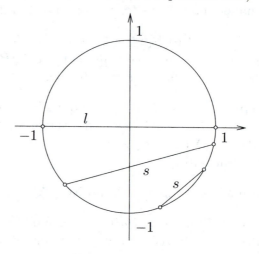

Figure A9

17. Let q, r be any two lines in $\mathbb{H}$ such that $q \cap r = \emptyset$ and the Euclidean lines containing q and r do not meet at a point of the unit circle $\mathcal{C}$ (see Figure A10). Prove that q and r have a common perpendicular line in $\mathbb{H}$. (Hint: Use Problems 14 and 16.)

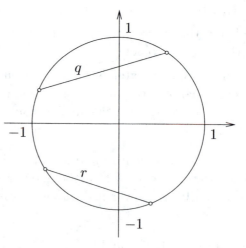

Figure A10

18. Let l be as in Problem 14 and let p be a line in $\mathbb{H}$ with Euclidean equation $y = nx + n$ for some $n \in \mathbb{R}$ (see Figure A11).

(a) Prove that l and p do not have a common perpendicular line in $\mathbb{H}$.

(b) Prove that l and p are asymptotic parallel lines in $\mathbb{H}$. That is, show that

$$\lim_{x \to -1+} d_{\mathbb{H}}((x, 0)(x, nx + n)) = 0 .$$

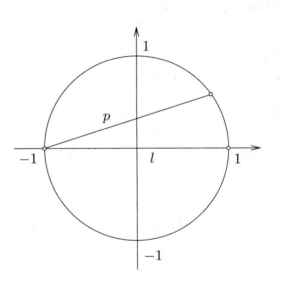

Figure A11

Bibliography

We list here only the books that directly influenced this text. In particular, the historical discussion in Chapter 0 owes a debt to [E], [R], [S], and [vdW]; most of the logic puzzles in Chapter 3 are taken from [W]; and our approach was inspired by [K].

[B]   Bonola, R., *Non-Euclidean Geometry*, Dover Publications, New York, 1955.

[G]   Greenberg, Marvin J., *Euclidean and Non-Euclidean Geometries*, 3rd edition, W. H. Freeman and Co., New York, 1993.

[E]   Eves, Howard, *A Survey of Geometry*, revised edition, Allyn and Bacon, Boston, 1972.

[H]   Heath, T. L., *The Thirteen Books of Euclid's Elements, Vol. I*, Dover Publications, New York, 1956.

[K]   Kay, David C., *College Geometry*, Holt, Rinehart and Winston, New York, 1969.

[M]   Martin, George E., *The Foundations of Geometry and the Non-Euclidean Plane*, Springer-Verlag, New York, 1975.

[PJ]   Prenowitz, Walter, and Jordan, Meyer, *Basic Concepts of Geometry*, Ardsley House Publishers, New York, 1989.

[R]   Rosenfeld, B. A., *A History of Non-Euclidean Geometry*, Springer-Verlag, New York, 1988.

[S]   Stahl, Saul, *The Poincaré Half-Plane*, Jones and Bartlett Publishers, Boston, 1993.

[vdW]   van der Waerden, B.L., *Geometry and Algebra in Ancient Civilizations*, Springer-Verlag, Berlin, 1983.

[WW]   Wallace, Edward C., and West, Stephen F., *Roads to Geometry*, 2nd edition, Prentice Hall, Upper Saddle River, 1998.

[W]   Wylie Jr., C. R., *101 Puzzles in Thought and Logic*, Dover Publications, New York, 1957.

Index

absolute plane, 55, 56, 143, 214
acute angle, 6, 127, 153, 168, 175, 219, 233, 263
Acute Angle Theorem, 219
additivity of defect, 215
Alexander the Great, 3
Alexandria, 3
all, 33–34
Almost-Uniqueness for Quadrichotomy Theorem, 94
alternate interior angles, 188, 190, 198
and, 32
angle, 3, 127, 262
    measure, 10, 19, 26, 27, 127, 143
    sum, 6, 141, 196, 200, 203, 209, 212, 214
Angle-Angle-Angle Criterion for Congruence, 246
Angle-Angle-Angle Theorem, 246
Angle-Angle-Side Theorem, 176
Angle-Angle-Side Almost Theorem, 206
Angle-Side-Angle Theorem, 147
angular distance, 115, 119–121, 124
antipode, 18, 93–94, 111
Antipode-on-Line Theorem, 93–94
aphelion, 244
arc, 18, 19, 121
    length, 19
    minor, 18, 68–69, 79

Archimedes, 4, 6
area, 1, 6, 12
Autolycus, 9
axiom system, 55, 143
axioms, 3, 11, 16, 55–63, 79, 81–83, 86–87, 110, 111, 115–118, 125–126, 142–146, 265
Axioms of Angular Distance, 115
Axioms of Distance, 56, 143
Axioms of Incidence, 57, 143

Babylonia, 1
Baudhayana Sulvasutra, 1
Beltrami, Eugenio, 8
between, 65, 116
betweenness, 11, 78, 124
    of points, 65
    of rays, 116
Betweenness of Points Axiom, 79, 143
Betweenness of Rays Axiom, 117, 144
birectangular triangle, 160–161
Birkhoff, George David, 11, 146
bisector
    of an angle, 128, 220
    perpendicular, 160, 217, 234
Bolyai, Farkas, 7–8
Bolyai, János, 7–10
bound
    greatest lower, 52–53
    least upper, 51
    lower, 52